T0316737

Listening in the Field

Inside Technology
edited by Wiebe E. Bijker, W. Bernard Carlson, and Trevor Pinch

A list of the series appears at the back of the book.

Listening in the Field

Recording and the Science of Birdsong

Joeri Bruyninckx

The MIT Press
Cambridge, Massachusetts
London, England

This book was set in ITC Stone Sans Std and ITC Stone Serif Std by Toppan Best-set Premedia Limited.

Library of Congress Cataloging-in-Publication Data

Names: Bruyninckx, Joeri, author.
Title: Listening in the field : recording and the science of birdsong / Joeri Bruyninckx.
Description: Cambridge, MA : The MIT Press, [2018] | Series: Inside technology | Includes bibliographical references and index.
Identifiers: LCCN 2017032849 | ISBN 9780262037624 (hardcover : alk. paper) ISBN 9780262553391 (paperback)
Subjects: LCSH: Birdsongs--Recording and reproducing. | Birds--Vocalizations.
Classification: LCC QL698.5 .B79 2018 | DDC 598.159/4--dc23 LC record available at https://lccn.loc.gov/2017032849

148702397

Contents

Acknowledgments

In researching and writing this book, I have accrued many debts of gratitude. My thanks go first of all to the many biologists, recordists, and archivists I have met along the way. They have welcomed me into their homes, gardens, and archives, dug up valuable documents, and taken the time to reflect on their work. In particular, Cheryl Tipp, Richard Ranft, the late Jeffery Boswall, Hans Slabbekoorn, Magnus Robb, Greg Budney and the staff of the Cornell Laboratory of Ornithology, Randy Little, Karl-Heinz Frommolt, the late Günter Tembrock as well as his wife Sylvia, and many other field recordists at the British Wildlife Sound Recording Society and the Dutch Club voor Natuurgeluid Registratie have helped guide my investigation. The staff at the BBC Written Archives Center, the Cambridge University Library, and the Cornell University Library have kindly helped me get access to key sources.

Mentors, colleagues, and friends have sustained this project in its many iterations. This book originated as a doctoral dissertation at the Faculty of Arts and Social Sciences of Maastricht University, where Wiebe Bijker and the teaching staff in the Cultures of Arts, Science and Technology program sensitized me to the many unexpected entanglements in this field. That this project came to further fruition, however, owes much to Karin Bijsterveld, who lent inspiration and direction to the project, but also shaped the contours of my scholarship and continued to support my work throughout. Dissertation committee members Jo Wachelder and Jens Lachmund have each shaped my thinking, both by example and with their keen and perceptive feedback. The ideas and direction in this book were further sharpened in exchanges with numerous peers and friends. My colleagues, old and new, in the Maastricht University STS colloquium, Summer Harvests (where researchers present work in progress), the FASOS Graduate School, and in particular the members of the Sonic Skills research group, Anna Harris, Stefan Krebs, Melissa van Drie, and Alexandra Supper, have made essential contributions as interlocutors, readers, and writing partners. They, together with Tim van der Heijden, Vincent Lagendijk, Caoilinn Hughes, Verena Anker, Fabian de Kloe, Eefje Cleophas, Bart van Oost, Ties van de Werff, Constance Sommerey and Koen Beumer, have offered valuable insights and pleasant diversions. Outside Maastricht, the Netherlands Graduate Research School of Science, Technology and Modern Culture (WMTC) has brought together a wonderful network of people in STS and extended my horizons. They, together with copanelists, discussants, and audiences at various colloquiums, workshops, and conferences, have impacted this work in unmistakable ways. I am also very grateful to Trevor Pinch, whose enthusiasm for the field has been inspiring. He and the faculty and graduate students in the Cornell STS department provided a most stimulating environment for me to work in. After I finished my doctorate, during a stay at the History and Anthropology of Science, Technology, and Society Program at MIT Rosalind Williams, Stefan Helmreich, and the rest of its staff and students helped me look at this project in new ways. All the while, the Maastricht Science, Technology and Society Studies research programme, headed by Harro van Lente, has continued to be a steady intellectual home port. In its final stages, this book has been completed at the Max Planck Institute for the History of Science, where Viktoria Tkaczyk brought together a wonderfully inspiring group of fellows as part of the research group Epistemes of Modern Acoustics. The group and the institute as a whole have provided a vibrant forum for exchange and the perfect place to finish this book. I am grateful for all of their support.

Kate Sturge has been fabulous at editing the entire manuscript. She has helped me shed unnecessary padding and own the rest. I would like to thank the anonymous reviewers in various stages for their enthusiastic, thorough, and thoughtful readings of the manuscript drafts. At the MIT Press, Margy Avery, Katie Helke, Laura Keeler, Kathleen Caruso, Elizabeth Judd, and the series editors have effectively guided this project to completion. Cynthia Landeen produced an excellent index. Oxford University Press and Sage Publishers kindly granted me permission to draw on arguments I developed originally in *The Oxford Handbook of Sound Studies* and *Social Studies of Science* but have reworked for this book. The Faculty of Arts and Social Sciences of Maastricht University and the Netherlands Organisation for Scientific Research (NWO) have generously funded part of the research that has gone into this book. In addition, NWO and the Max Planck Institute for the History of Science have helped to ensure that the book is appropriately illustrated.

Finally, family and friends have brought perspective to the writing of this book. My parents, Noortje and Robby, Dirk and Josie, have been great advocates and a source of many animated discussions on anything but birdsong. Marijke, finally, has been there all along. She has been an inexhaustible source of love, support, and common sense. Lucie arrived on the scene at the very end and as I complete these pages, she has already started to imitate the animals she encounters. Her coos and moos remind me of all those things that cannot be recorded. This book is for both of them.

1 Eavesdropping in the Wild

A Faint Cry

On April 9, 1935, a group of five ornithologists perched on a trunk deep in the heart of the "Singer Refuge," an area of swampland near Tallulah, Louisiana. The party had stopped during a fifteen-thousand-mile round-trip expedition from Upstate New York to study and collect the voices of American birds. Funded by a former Wall Street investment banker turned ornithologist, Albert Brand, and supported by the American Museum of Natural History and Cornell University, the party planned to document the lives of birds on the brink of extinction. Reports had reached them that the crowning glory of American birdlife, the elusive ivory-billed woodpecker, had been spotted in these marshes. Somewhat more than a decade earlier, the head of the expedition, leading field ornithologist Arthur A. "Doc" Allen, had witnessed a pair of ivorybills singing, but by now advancing loggers and trigger-happy collectors among his fellow ornithologists had made such sights increasingly unusual.[1]

For several days, the party had waded through flooded bayous and muddy virgin forest. When they finally discovered a nest, they hauled in two truckloads of equipment to record the birds' behavior on photographic plates and sound recording disks. They had been perfecting their technique in the woods around the Cornell campus for some years, but this kind of wilderness—remote, dirty, and humid—put it to a harsh test. Sitting in their improvised studio for a week, bent over field notes and eavesdropping on the woodpeckers' calls through earphones, they created a set of recordings that would be the species' first and only ones (see figure 1.1). By midcentury, sightings had become extremely rare and the bird was generally presumed extinct.

That changed seventy years later, almost to the day, when the Cornell Laboratory of Ornithology and the Nature Conservancy jointly published

Figure 1.1
Cornell ornithologist Peter P. Kellogg on a field expedition, in an improvised studio observing and recording a nest of ivory-billed woodpeckers, in 1935.
Source: Albert R. Brand Papers, #21-18-899, Folder 2:1, Division of Rare and Manuscript Collections, Carl A. Kroch Library, Cornell University.

a paper in *Science* announcing the rediscovery of the species nicknamed "ghost bird" (Fitzpatrick et al. 2005; Steinberg 2008). Their evidence included a number of sightings and some grainy video material. But with visibility severely limited in the densely forested tracts of Arkansas and Louisiana, the researchers gave special prominence to a series of brief recordings that captured what sounded like the birds' distinctive double-knock drum and nasal tooting calls. As the only acoustic reference point for the ivory-billed woodpecker, the historic sound recordings of 1935 now assumed a key role in assessing the evidence for the bird's alleged rediscovery. Mounting new expeditions, the search teams used the recordings to train volunteers and help analysts verify new registrations of possible ivorybills, aided by a programmed pattern-recognition algorithm that scanned hours of automated recordings.

Much was at stake in the validation of these claims of rediscovery, and not just for the ornithologists and academic institutions whose hard-earned reputations were knitted into the evidence they endorsed. Over the course of the twentieth century, the ivory-billed woodpecker had become a veritable icon of the American wilderness. A year after the historic sound recordings had been made, the influential American conservationist Aldo Leopold (1936) declared this "Lord God Bird" and "King of Woodpeckers" an inextricable part of the American pioneer tradition, representing a vanishing inner frontier fundamental to the American national character.[2] A symbol of loss, its presumptive rediscovery also promised a second chance for nature conservation; incontrovertible confirmation of the species' continued existence would allow conservationists and policymakers to put forward a concrete case for the legal protection and preservation of a large area of land in the southeastern states. The audio, both historic and new, effectively accentuated this emblematic value. Bird vocalizations have been powerful icons in conservation discourse, nourishing hope and affect from Rachel Carson's "silent spring" back to the late nineteenth-century environmental movements that rallied middle-class bird enthusiasts to the cause of wildlife protection. In this discourse, birds' calls and song served to suggest rapprochement with the natural world and to make the impact of human activity on ecology tangible. Tapping into this symbolism, the specter of an ivorybill calling from across the brink of possible extinction carried deep into the popular imagination.

But however strong its symbolic currency, the recorded evidence also caused a rift in the ornithological community.[3] A vocal group of skeptical ornithologists considered the new recordings too faint, too brief, or too masked by noise to permit conclusive proof of the bird's existence. They countered the visual and acoustic evidence with their own analyses, mobilizing amplitude displays and audiospectrograms to suggest that the recorded sounds could also be explained by nearby traffic, gunfire, other bird species with similar calls, or even other observers who were mimicking the famous "kent" calls with instrument mouthpieces or broadcasting the historic Cornell recordings in an attempt to provoke a response from hidden woodpeckers. A highly charged trial of testimonials and rebuttals ensued, buttressed by a series of lengthy, expensive, and much-publicized search expeditions. But although ornithologists continue to roam the Singer Refuge today, their searches have not yielded the incontrovertible proof many had hoped for. As the controversy faded from public view, it left the ivory-billed woodpecker suspended between survival and extinction.

What does it take for an ephemeral sound, a faint cry, to be registered under the unsympathetic circumstances of field research, and to be recognized as an object of scientific evidence? How did sound recording come to be considered an authoritative and reliable way of studying wildlife, and why does it continue to be contested nonetheless? And how did the very act of recording animal voices in their own environment change how scientists, and ultimately the public at large, came to listen to them? *Listening in the Field* draws out answers to these questions by tracing a history of the practice of sound recording in the field and its role in the development of birdsong biology. Turning to the analog era of recording between 1880 and 1980, it reconstructs the material and cultural conditions that have allowed sound registrations of wildlife and birds in particular to be crafted, appreciated, and contested—as objects of both scientific investigation and popular fascination.

The practice of recording birdsong took shape at a peculiar intersection of turn-of-the-twentieth-century popular entertainment, pedagogy, and field ornithology. But in the course of that century, ornithologists, taxonomists, ecologists, and ethologists transformed what had originated as a naturalist fascination with the music-like attractions of living birds into a scientific approach to the study of animal life. In the wake of their discipline's specialization, biologists of diverse plumage turned to new media and communication technologies to capture and measure sounds for their own specific purposes.

Taxonomists have used sound recordings to investigate species relations, alongside more traditional markers of distinction such as feather coloration or anatomy, resulting in the discovery of new species (Alström and Ranft 2003; Hardy 1969). Ecologists have deployed sound recording and analysis as a method for tracking, monitoring, and counting species where conventional techniques fail—in hard-to-survey areas such as densely forested tracts or during nightly migrations (Baptista and Gaunt 1997). Ethologists have developed sound recording into a technique for investigating various aspects of birds' singing behavior, as an index of how birds learn, inherit, or vary their songs or—by playing back sounds to elicit a vocal reaction—their social functions (Marler and Slabbekoorn 2004). By midcentury, such diverse work had begun to converge in a new interdisciplinary field under the heading of "biological acoustics," which elevated advanced techniques of acoustic analysis to the status of a methodology for exploring fundamental mechanisms of instinct, behavior, and evolution. Scientists' investigations were further supported by sound archives across North America, Europe, and elsewhere, which collectively made hundreds of thousands of recordings available for bioacoustic research, analogous to the collections of physical specimens such as eggs, skins, or mounted exemplars that exist in other branches of natural history investigation.[4] In several respects, then, sound recordings became the standard tool to document, study, and preserve the vocal behavior of birds.

But as the ivory-billed woodpecker's vexed appearances suggest, even in the twenty-first century the evidentiary value of recorded sound has remained far from undisputed. In fact, the controversy punctuates a far more messy and ruptured history of methodological innovation, in which the fervent instrumentalization of new media technologies in wildlife research gave rise to a long-running and pervasive debate among field biologists regarding the conditions of fieldwork, the calibration of sensory judgment, the epistemic qualities of sound and listening, and so ultimately, the nature of scientific evidence.

In the course of their academic professionalization, ornithologists had learned to trust the tangible presence of captured specimens or—enabled by technologies for extending and disciplining observation, such as binoculars and field guides—a visual identification or observation by peers that could be trusted to be reliable (Barrow 1998; Dunlap 2011; MacDonald 2002). Sound problematized these evidencing practices. Ornithologists typically relied on auditory cues to guide their work in the field, for instance, when they listened to locate a potential specimen in the field. But transforming such (often tacit) intuitions into a category of proof required a new kind of private experience to be made accessible to a collective of scholars. For listening to be crafted into a form of scientific observation and data collection, new techniques of description and evidencing had to be contrived, new ways devised for listeners to learn and to convey their findings, and new routines cultivated for calibrating and assessing the reliability of their audition. It also required listeners to define what exactly they were listening to.

New sound technologies occasioned and, at least at first glance, also seemed to resolve such questions. But as this book shows, they actually greatly complicated such issues. For although sound recording technologies promised to relieve listeners by registering, calibrating, and preserving sound for them, they also compelled scientists to engage with a set of practices that were never exclusively scientific nor originally theirs to define. For sure, many early recording devices had originated as experimental or documentary tools, yet in most cases this scientific use had long been eclipsed by the cultural and commercial appeal of recorded music (Kursell 2012; Morton 2004). It is easy to imagine ornithologists' attraction to a technological medium whose promise was anchored in a discourse of fidelity and the possibility to transport sound reliably across time and space. But that promise had been defined as much by the needs of advertising, mass communication, and military applications, as by scientific aspirations. When biologists turned to record birdsong in the field, they thus appropriated a class of media technologies whose conditions of use had crystallized within settings as diverse as musical composition, nature teaching, radio broadcasting, gramophone entertainment, hobby electronics, and Cold War surveillance.

It is this tension—between sound recording as a cultural practice and its appropriation as a distinctively scientific technique, between sound as a form of evidence and its cultural existence—that propels the story in this book. In redefining sound recording into a methodology of scientific investigation, these biologists could not avoid grappling with those various conditions of use. First, making sound recording work in the field required specific resources and expertise that had accumulated over time in these existing milieus, such as infrastructure, equipment, funding opportunities, and networks of communication, along with specialized "sonic skills"— listening skills and techniques to make effective use of the equipment (Supper and Bijsterveld 2015). Second, even as scientific objects, recordings of animal vocalizations retained a broad-ranging commercial, artistic, educational, and sometimes even military appeal. Finally, vernacular meanings and practices of recording continued to pattern how people heard and appreciated recorded sound. Hence, materially, socially, or discursively, sound recording never existed exclusively as scientific practice. Nor was it, however, meant to. In developing and legitimizing sound recording as a scientific technique, scientists used these conditions to their benefit. *Listening in the Field* argues that when and where sound recording thrived, it did so precisely because ornithologists sought to exploit these links and often managed to repurpose sound recordings beyond the direct circle of their peers who were invested in birdsong. Sound recordings were actively collected, processed, and exchanged with a broader set of purposes in mind than an exclusive use as scientific evidence; as wildlife sound gramophone recordings, radio broadcasts, movie soundtracks, musical compositions, or didactic resources, they traveled back and forth between scientific and popular domains. It is this entanglement of purposes that has helped tie together a broad-based community of listeners in the field, but it has also led them to listen to these recordings and their natural environment in very different ways.

Scientific Recordings in the Field

If we are to gain a better understanding of how listening and sound recording actually acquired epistemic currency in field science, we must turn our focus to their material existence and the specific social, cultural, and technological contexts within which they thrived. To do so, I draw on several recent strands of scholarship in the history of science, technology, and the environment, in combination with a broader cultural history of the senses and experience. Indeed, this book aligns most conspicuously with a recent concerted effort to explore the various intersections between the extensive realms of science and technology and those of sound. Such work has pushed to detail among other things how acoustic knowledge and its derivatives in the science and technology of physics, architecture, psychophysics, and communication engineering have helped to articulate—and were themselves prompted by—broader cultural conceptions of aesthetics, music, signal, and noise.[5] Such work has also begun to consider the historical formations of listening itself to show that outside the concert hall, the ear has been deployed as an epistemic tool and mode of knowledge acquisition in settings that range from car repair and telegraphy to high-energy physics.[6] It has shown that listening, as a cultivated practice associated with knowing, has emerged and prospered under specific historical conditions.[7]

This book is part of that effort, tracing over a period of a century the formation of listening as a practice of investigation along with the boundaries of the field and the kinds of knowledge it helped produce. But in doing so, it also historicizes listening practices in a new way, by situating their development in a much broader history of scientific observation. Listening has historically been parsed into distinctive modes of attention. Just as it has evolved to suit the needs of the physician diagnosing an illness or the engineer monitoring technical processes, scientists have calibrated listening to explore the world for new phenomena, and to categorize or analyze them in detail. Examining the kinds of knowledge and professional status that these listeners have claimed contributes to understanding how the world has been heard differently. But it also expands an understanding of scientific practice itself. While observation is routinely cast in visual terms, listening too has involved the sharpening of attention, the disciplining of experience, the calibration of judgment, and the formation of "thought collectives."[8] Historians and sociologists of science and technology have slowly but surely acknowledged the importance of physical presence and full-bodied experience, and this book uses field recording as a case to work through some of these themes.[9] Ornithologists and other field scientists have traditionally been very active in developing modern techniques of observation, by means of field guides, systems of recordkeeping, or specialized forms of reasoning.[10] But in making birdsong perceptible, definable, and knowable, ornithologists have likewise had to develop skills, concepts, and languages for talking about sound—by inventing them anew or by appropriating them from listening cultures in music, linguistics, or communication engineering.

The history of scientific listening in ornithology is intimately entwined with its practices of recording. Representations, after all, serve to structure observations for others to access. Although sound recording has conventionally been framed within a cultural and socioeconomic history of twentieth-century mass entertainment, its development has frequently intersected with investigatory practices in such fields as anthropology, linguistics, zoology, and comparative musicology.[11] Recent histories of these fields underline the importance of sound recording technologies for their development.[12] Once initially elusive and transient objects of study such as voice, language, or music could be reproduced across time and space, they could be incorporated into the archival, documentary, and experimental practices that have defined the projects of nineteenth- and twentieth-century scientific investigation. As a reconstruction of how this has changed field biology, my story resonates with existing case studies. But it also articulates a more specific set of issues.

With few exceptions, practices of sound recording in science are subsumed under an intellectual history as an incidental technical precondition that triggered or sometimes resolved substantive controversies.[13] However, as we will see, sound recordings themselves did not remain uncontroversial and their methodological legitimacy has often been closely tied to their cultural and material lives. Existing histories of scientific sound recording typically tend to situate the methodological legitimization of sound recording within a discourse of early twentieth-century mechanization and instrumental mediation that historians of science Lorraine Daston and Peter Galison (2007) have termed *mechanical objectivity*, a discourse that threads together the phonograph with the advent of daguerreotypes, cameras, and graphic recording instruments in the life sciences.[14] According to this powerful and pervasive discourse, the listening subject was to be ousted by the mechanical ears of the phonograph. Instead of relying on listening skills and aural experiences, the phonograph and other types of sound recording promised a registration untainted by the personal tastes, fallible judgments, and cultural biases of a human observer. This discourse sets an important reference point for understanding why and how sound recording technologies achieved acclaim in the diverse fields I have listed so far, and we will see it being mobilized on several occasions, alongside other such promises. Sound recording promised a more fine-grained type of analysis, because it enabled sounds to be repeated, measured in different ways, to be visualized and compared as formal patterns. It also allowed researchers to tune into frequencies not usually audible to human hearing. These perceived advantages go a long way in explaining why these instruments achieved the currency they did. But at the same time, a pervasive narrative of users' ostentatious confidence in their epistemic value needs to be complemented with a story that is far messier and far more conflict-ridden. It requires attending to technologies' existence not just as deep-seated textual identities, but also as embodiments of alternative forms and meanings.

As Lisa Gitelman points out, media technologies tend to sink to a point of apparent transparency, where users become oblivious of the protocols and norms that they heed, in favor of the phenomena and information that such instruments are said to mediate. But when recording technologies in the widest sense of the word, such as paper diagrams, electric microphones, portable magnetic tape recorders, or sound spectrograms, are brought into a new context, handled by new kinds of users, applied to new questions, or confronted with new practical problems and sounds, they expose themselves as sites of ongoing negotiation. As the ivory-billed woodpecker case suggests, a technology's protocol may be breached or questioned at any time. But it is during what Gitelman (2008, 1) calls their "novelty years, transitional states, and identity crises" that technology and practice mutually redefine each other most conspicuously. Such changes may become visible as physical alterations, shifting meanings for different users, or be expressed in the tacit skills and embodied choreographies of users and their tools. Eventually they sink in to identify with the technology's new deep-seated assumptions.

Tracing sound recording's development into a medium for scientific listening therefore requires picking apart the category of "recording" itself into the bundle of inscriptive media that it represents, each made up of different kinds of hardware, structured according to specific protocols and formats and invested with very different meanings and functions. Here I take a cue from social studies of scientific imaging and visualization, a strand of science and technology studies that investigates forms of representation and their specific roles in knowledge production.[15] Since the 1980s, this field has brought forth a rich vein of ethnographic, historical, and philosophical analyses of the visualization technologies—sketches, images, graphs, tables, and scans to computer screens and simulation models— that are inextricably woven into the fabric of scientific practice. Attending closely to the embodied and situated nature of the people, tools, and objects involved in "representing," this work has shown the production and use of visualizations to be fraught with material resistances, tacit knowledge, and negotiations of trust, authority, and expertise. This materiality not only complicates representations' production and circulation, it conditions the epistemological claims and ontological orderings they support.

First, like other forms of representations, sound recordings are bound by their materiality. Sounds may be sketched in pencil-drawn diagrams, traced by a stylus in warm wax, arranged as magnetic particles on tape, or transferred as electronic bits. Such material forms matter because they display specific sensitivities to acoustic phenomena. It makes a difference whether a vibration resides in the wavy line of a time amplitude or a microtone pitch scale, whether it is played back from the wavering reel of magnetic tape or discerned in the thick-brushed ink blob etched into a sheet of conductive sonogram paper. These forms' approaches to storing, transforming, manipulating, and reproducing sound structured the contents and practices of birdsong biology in a myriad of ways. Recorded on different media or in different formats, sounds became more or less durable, labor-intensive to reproduce, intuitively legible, easy to listen to or to inspect visually. But they also came to be understood to exist within very specific limitations of time or frequency range, manifesting their dynamic range over minute variations, emphasizing an expressive timbre over a surprising flourish.

Second, these different forms matter because they were conducive to broader social, cultural, and economic conditions. Recording media can be more or less easy, cheap, quick, desirable, marketable, or reliable, to make and subsequently to duplicate or circulate. Recording technologies require a level of skill, knowledge, and resources that birdsong biologists and their collaborators sometimes possessed, to a greater or lesser extent. These became the conditions under which dependencies and collaborative networks were forged. Such conditions shifted, moreover, over time. The expense and complexity of gramophone recording led field ornithologists to forge alliances with electrical engineers and recording technicians, just as hobby recordists developed a dependence on field ornithologists to produce recordings of sufficient quality to achieve the recognition they sought. On the other hand, some of these media were inexpensive enough for small groups of dedicated university scientists or even individual recordists to acquire or build themselves, to make adjustments, and to disseminate significant innovations among their peers and, occasionally, to convince instrument manufacturers of their worth. As such, these forms also came to affect in part which users participated in the production of wildlife sound recordings, how, and with what degree of success—scientific, commercial, or otherwise.

That success, however, was largely defined against the conditions of twentieth-century field research. Historians of science have recently begun to address scientific work in the field, and the history of scientific sound recording folds into many of the social and cultural transformations through which the field site emerged as a place of scientific investigation.[16] When ornithology became established as a specialized field of study in the middle of the nineteenth century, its province was largely confined to museum collections of mounted specimens. The specimens were being shot and collected in the field, but authoritative academic work itself focused on ordering taxonomic specimens indoors. By the end of that century, however, shifting conceptions of nature and a growing popularity of birdwatching led large groups outdoors to observe animals in the field. Yet when field observation had become a widespread practice, the field site continued to be considered epistemically ambivalent. Unlike natural history museums, laboratories, botanical gardens, or even zoos, the ornithological field site—in nature reserves, forests, beaches, or city parks—could never be cordoned off from the places in which it was established. Indeed, that emplacement was exactly the point, for it allowed nature to be observed and studied in all its interaction with the environment, in all its particularities and variability. But phenomena in the field were not easy to control or replicate, which made them suspect in comparison with the laboratory's trust-inspiring rhetoric of placelessness that ensured its virtual ubiquity in the scientific landscape from the mid-nineteenth century onward. When considered against their claims to the laboratory's authority of universality and control, field studies appeared messy and uncontainable. In response to this ambiguity, some field scientists strove to bring elements of the laboratory into the field, while others chose to assert the value of place-based research as distinct from the artificial environment of the lab (De Bont 2015; Kohler 2002b).

As a natural place and public space, moreover, the field was not just physically accessible to professionally sanctioned scientists. Rather, it was shared with hunters, tourists, commuters, or laborers, making encounters with (and often practical reliance on) such local passersby and their social activities quasi-inevitable. This physical accessibility was one of the factors that granted field studies of birdsong their broad-based attention. In comparison to other natural subjects, birds are a widely and generally distributed, highly audible and visible group of animals, which makes them relatively easy to encounter and spot in the wild. Moreover, traditional designation of their vocalizations as "language" or "songs" and description in aesthetic terms have contributed to their strong cultural resonance. These factors have helped to ensure that the scientific work that I conveniently collect here under the rubric "birdsong biology" actually fans out into a wide range of fieldworker identities. These include academic professionals, ornithologists next to newly emerging specialists such as ethologists, ecologists, systematists, and bioacousticians, as well as a diverse group of "amateurs," dedicated naturalists, recreational birdwatchers, or civic enthusiasts whose influence in the field sciences, particularly ornithology, has been pervasive and vital to their development (Ainley 1979; De Bont 2015; Mayfield 1979).

But in addition, other, often unsuspected groups of actors have made their mark on this field. As recent scholarship in the history of animal science and conservation suggests, animals' cognitive, expressive, and musical capabilities have not only preoccupied (amateur or professional) scientists; they have also been assured of occasional attention from the military (for instance, in naval bioacoustics) and countercultural or new wave movements.[17] Extending that perspective, my account recovers the contributions of an extremely varied group to the biology of birdsong. When biologists turned to record the sounds of birds, they were accompanied by hobby recordists, amateur sound hunters, nature teachers, musicians and composers, broadcasting icons, movie producers, and electrical engineers, operators, and technical assistants. These groups recorded bird vocalizations for their own intellectual, aesthetic, or economic motives, yet their skills, resources, and expertise were crucial to the development of scientific field recording.

Biologists did not just share the field site with these groups. They actually relied on them for the production and exchange of field recordings. Their shared preoccupation with birdsong, and the expense, effort, and technical and logistical complexity of field recording, tied them into tenuous networks that rendered the boundaries between scientific and other kinds of wildlife recording flexible and dynamic. When the British Broadcasting Corporation (BBC) extended its sound archive of wildlife sounds for use in its programming in the early 1950s, for instance, it hired an amateur ornithologist to run the archive. Recordings were sourced from amateur sound hunters, professional wildlife recordists, and radio organizations. The BBC traded such recordings with other international sound archives and donated a copy of its archive to scientific or semiscientific naturalist organizations and academic laboratories in Britain. Those archives were subsequently used to liven up popular lectures, but also to pioneer new analytic techniques and develop new hypotheses. Likewise, when Cornell ornithologists were recruited in the 1940s to record animal voices in the

Central American jungle and investigate its acoustic properties to prepare military personnel operating in the Pacific region, their recordings were reused for release on gramophone.

At the points of intersection between a set of relatively accessible new media technologies and a newly forming research field, the shared preoccupation with birdsong often blurred the distinctions between technical and scientific expertise and forged new identities. Bioacousticians co-opted, for instance, a technical culture of electronic tinkering, and enlisted recording enthusiasts and hobbyist audiophiles as both producers and consumers of field recordings and technical innovations in recording equipment.[18] Celebrating outdoor recording as a pastime that married the joys of being outdoors with the wonder of natural history work and the technical challenge of field acoustics, the identity of "sound hunter" provided scientists with an eager field collaborator and a reliable collector of acoustic specimens.[19] Scientists, too, adopted a more explicitly hands-on and technical identity. Interacting with recording technicians, electrical engineers, and instrument manufacturers, they built up the technical expertise that allowed them to influence the direction of the field. As instrument users, for instance, they adapted existing instruments and techniques to their own needs and abilities, prompting various technical innovations that also structured the community of field recordists. Amateur ornithologists developed new graphic sound notations for nonmusical, natural sounds, while bioacousticians built and redesigned recording equipment and rewired sound spectrographs, sometimes to the approval of instrument manufacturers, who incorporated their tweaks into marketable applications.

These crossovers between various kinds of field recordists allowed scientists to take an active role in giving shape to their community and the practice of scientific field recording. They not only pursued their own scientific direction, but as entrepreneurs also ventured into the various contexts that sustained their work, such as the entertainment industry. When the Cornell and American Natural History Museum recordists set out on their recording expedition in 1935, they already anticipated that the results would be beneficial not just to scientific study. The outcomes were also used to add luster to exhibitions and taxonomic displays at both institutes; they were marketed as soundbooks and gramophone recordings, for didactic purposes in the classroom, to be enjoyed in the domestic sphere, or to be used by aspiring birdwatchers learning to recognize common birdsongs. Recordings were also used as sound effects: in local and national wireless broadcasts and, with the advent of sound motion pictures, on soundtracks of movie productions, nature films, and television broadcasts.

Recent scholarship on the historical intersections of science and the entertainment industry has highlighted the role that mass communication media such as radio, television, and documentary and fiction film have played in people's learning about science, outside schoolrooms and textbooks.[20] Such work has situated these media in a continuous field of tension defined by the need to be both dramatically entertaining and scientifically accurate. This book shows wildlife recording to be subject to similar considerations. That meant reconfiguring the procedures and aesthetics of a studio art conventionally focused on sound fidelity into a scientific research practice bearing the requisite marks of objectivity and authenticity. It also involved redefining the ears of the scientists' collaborators and listeners, accustomed by a recording industry invested in animal imitations, to appreciate a different kind of realism of animal sounds recorded under the difficult acoustic conditions of the field.

But if current scholarship described scientists mostly in an assistive, advisory role, as contributors of content or to verify its accuracy and logic, my account shows that sound recordings allowed scientists to carve a different role for themselves. By licensing and selling their own sound recordings, scientists could leverage their participation in these communities; it helped them to marshal resources, to co-opt skilled collaborators, and to extend a community of field recordists and data collectors. It enabled scientists, moreover, to engage not just in field recording itself, but to foster a technical culture of field recording and to circumscribe its accepted meanings, uses, and values. Sound recordings helped them do just that, by inspiring users to listen in particular ways, providing technical identities, standardizing discourses, or setting technical benchmarks for good recording.

But although scientists' entrepreneurial approach to sound recordings helped them to expand and consolidate bioacoustic research and reach out to the communities that sustained them, it also required them to engage with rules and norms for very different domains. Sound recordings that had been produced by scientific recordists, the entertainment industry, or citizen and amateur collaborators were each nested in very different domains, with their own (often institutionalized) regimes of ownership and authorship. Exchanging and capitalizing such sound recordings required such regimes to be aligned, which raised for scientists larger questions with regard to the organization and the governance of the knowledge commons: to what extent should such recordings and the voices of wildlife that they represent be considered part of a knowledge commons, a domain of knowledge and information that can be freely shared among participants? Who owns these recordings; who is allowed to circulate them and to whom; and who benefits from them? In answering these questions, scientists and their collaborators strategically defined the boundaries of a community of field recordists, as well as a set of norms and rules of exchange.

Case Studies

It is not my intention in this book to present a comprehensive history of sound recording in birdsong biology, let alone biological acoustics more generally. Instead, my account focuses on a selection of small communities or networks around key individuals and institutions that allow me to analyze the conditions under which particular approaches to the study of animal vocalizations have emerged and became standardized. Geographically, my scope is limited to developments in North America, Britain, and to a lesser extent Germany. Although the study of birds has arguably been a global phenomenon, historians of science have generally attributed a pivotal role to these countries in the establishment of ornithological knowledge, certainly since Darwin.[21] The ornithological communities within these countries have been particularly strong forces in developing bioacoustic methods. This is not to say that studies of animal vocalization failed to achieve wider attention elsewhere. A transnational history of this field remains to be written, but it is clear that by the second half of the twentieth century, centers of bioacoustic research were emerging in France, Scandinavia, and the USSR.[22] In these regions, too, amateur recordists and radio services contributed to the collection of animal vocalization recordings even before World War II.[23] Such parallels provide a significant backstory to my focus on the extensive and more densely populated histories of scientific field recording in North America, Britain, and Germany.

One indication of this is the establishment of publicly and privately funded sound archives in these countries. Though not always created specifically as repositories for scientific research, such institutions acted as a collection point for recorded wildlife sounds and thus fostered the exchange and circulation of recordings between scientific and other cultural contexts, such as broadcasting, amateur recording, or art. Although nature sound archives proliferated across the globe in the 1960s and especially the 1970s, the Macaulay Library at Cornell University in Ithaca, New York, and the British Library of Wildlife Sounds Archive in London were arguably the first, largest, and fastest-growing archives of wildlife recordings worldwide (Ranft 2004).[24] These libraries have served as nodes in the various wide-ranging and usually informal networks of individual amateur recordists, commercial and public institutions, academic departments, and scientific laboratories that enabled the circulation and exchange technologies, recordings, skills, or ideas. Tracing the personal and institutional connections that facilitated (or impeded) such circulation, I have narrowed my focus to a selection of subnetworks of field recordists, naturalists, and bioacousticians that took shape around the Cornell Laboratory of Ornithology, the British Broadcasting Corporation, the British Institute for Recorded Sound (later the British Library Sound Archive), and the Cambridge University Department of Zoology outpost at its Madingley Ornithological Field Station. As prolific producers, collectors, and consumers of wildlife recordings, these institutes and the networks they helped to feed were interconnected. As trailblazing developers of techniques of sound recording and analysis that have been widely adopted in the biology of birdsong and beyond, they were key to bioacoustics' history.

I bring a specific transnational perspective to these localities, although I do not present them as cases within a strictly comparative matrix, nor do I consider their entanglements as specifically representative of a story of scientific internationalism and successful transnational coordination. Both approaches to the history of science and technology have recently proved particularly successful.[25] But while keeping sight of the national contexts that enabled a public broadcaster such as the British Broadcasting Corporation, for instance, to take up a key role in the development of field recording, my concern with how sound recording developed into a scientific methodology calls for a perspective that shows how sound recordings have traveled between field sites, laboratories, archives, and readers—thus crossing geographic and social boundaries that were not always specifically national. Failures among bioacousticians and sound archives to collaborate internationally could be due as much to differences in institutional policies, personal goals, or preferences for dealing with a particular medium as they were to different national contexts.

My approach is transnational, therefore, in puzzling out an account that—in the words of historian of science Jim Secord (2004, 668)—"keeps the virtues of the local but operates at a unit of analysis larger than a single country." Secord advocates this approach as part of a historiographic program for understanding knowledge as communication by focusing on its circulation among different publics. Such a program is well suited to studying how sound recordings and recording technologies have traveled between dispersed locales and how they have stabilized there as epistemically meaningful carriers of information. In placing emphasis on processes of translation, exchange, and transmission, this has meant attending to the local conditions of making and interpreting sound recordings and to actors' efforts to build and maintain the personal, regional, and occasionally transatlantic connections that facilitated those recordings' circulation. It also involves attending closely to the technologies with which the sounds were captured, stored, and decoded. As "social, cultural, and material processes crystallized into mechanisms," such technologies typically came to embody the convictions of their developers, users, and audiences (Sterne 2003, 8).

My empirical basis is a qualitative analysis of a broad corpus of historical sources. The primary data considered here consists, first, of academic and popular publications, especially a diachronic survey of articles in nine journals featuring work relevant to the biology of birdsong, broadly defined, since 1880.[26] As historians of biology have shown, the close reading of developments in the content and style of scientific journals over a lengthy period reveals the processes by which new ideas and methods enter an intellectual community (Battalio 1998; Nyhart 1991; Johnson 2004; Futuyama 1986). The journals were selected to represent relevant developments in the field, and include both key ornithological journals and more recent publications at the interface of ethology and bioacoustics.[27] I complemented my analysis with a selection of relevant field guides and handbooks. Handbooks proved especially useful for tracing changes in the state of the art in birdsong biology, while field guides, recording manuals, and songbooks provided insight into changes in the organization of observation and recording in the field and the habitus of the recordist.

Second, I consulted a number of unpublished sources, held at institutions such as the archives of the British Library Sound Archive and its wildlife section, the Humboldt University's Tierstimmenarchiv, Cambridge University Library, the BBC Written Archives Centre, Cornell University Library, and the Cornell Laboratory of Ornithology.[28] This body of primary documents includes correspondence (between scientists, archivists, amateur recordists, technicians, recording engineers, and broadcasters), diaries, manuscripts, leaflets, news clippings, field and laboratory records, editing forms, administrative materials, and the recordings themselves. These sources combined offered a thick trace of recordists' ongoing activities, concerns, problems, and solutions. Wherever available, I supplemented them with popular publications and biographical or autobiographical sources on the actors involved. Such accounts are often more stylized than the primary documents; they typically revolve around the motif of scientific heroism, in which the individual scientist overcomes the challenges set upon him by nature.[29] Yet when contrasted with publicly undisclosed sources, these items provide helpful insights into the prevailing scientific habitus and customs of field recordists.

I have also drawn on a selection of oral history interviews with record-ists, sound archivists, and bioacousticians. Some of these were collected by archival institutions as part of oral history programs on the develop-ment of field recording; others were carried out myself. My own interviews were designed to elicit interviewees' experiences of contemporary and his-torical recording practices, and often included participatory observation of historic recording equipment and techniques in operation. Although I do not base this account in any systematic way on ethnographic participant observation, I visited the field with amateur recordists and bioacousticians on several occasions. These observations provided a substrate of practical knowledge and material sensitivity that have helped me to interpret the written technical or biographical accounts.

Structure of the Book

The five chapters that follow offer a thematic account of the development of sound recording into a scientific technique. In the century between the first incarnations of mechanical sound reproduction in the 1880s and digital audio in the 1980s, students of birdsong appropriated different recording technologies. The chapters of this book are organized around the transitional periods and novelty years in which such technologies were tested, settled in, and became established. Particular technologies, such as the magnetic tape recorder and the sound spectrograph, often reinforced each other in the same period, which accounts for some chronological overlaps between chapters 4 and 5. This diachronic thread gives an insight into the historical development of the field of birdsong biology, its inhabit-ants, and its listening techniques. Each chapter, however, also illuminates these chronological developments from a particular vantage point, taking a technological innovation as a lens to explore the specific material, social, and representational technologies that have enabled sound to be repro-duced as a scientific object.

Chapter 2 focuses on a half-century period between 1880 and 1930, when birdsong was recorded in a medley of verbal descriptions, musical notations, and onomatopoeic syllables. As field observation gradually grew from a province of naturalist writing to a practice that concerned both recreational birdwatchers and professional ornithologists, these formats became topics of debate. In this chapter I trace a dispersed group of ama-teur naturalists and newly professionalizing ornithologists who sought to appropriate musical notation from an instrument of musical composition, nature teaching, and artistic performance into a workable and authoritative scientific technique in field observation. This, I show, was a complex process in which field observers argued to bolster their own epistemic authority but also to ensure a wide readership. Scientific listening thus took shape between two opposing pressures: on the one hand, an attempt to standardize birdsong recordings within a single, recognizably scientific format, on the other hand, an attempt to integrate the multiple interests, competencies, and listening techniques that ornithologists brought to the field. To understand how such formats came into scientific use, it is therefore important to consider that their use was never fixed. They were not just read as mimetic representations; as records of birdsong, these formats also acquired their meaning from their use in teaching, learning, and remembering such songs, and even in luring listeners to listen to them in other, unsuspected ways. Even though recording formats themselves changed, these functions would go on to structure how scientists and others engaged with birdsong.

As I show in chapter 3, recordings produced with electrical equipment integrated all four of those functions when they were first successfully produced in the field from 1930 onward. Moreover, they promised to replace inherently subjective impressions with objective registrations. This change would make specialized listening skills redundant, but it also restricted field recording to those with access to expensive and technically demanding equipment. Between 1930 and 1950, small groups of ornithologists in Britain and the United States ventured into field recording through various alliances with the recording industry. These gramophones and microphones came with their own practical and epistemological problems. Analyzing the problems of recording in the field and recordists' solutions to them, chapter 3 shows that sound recordings were artfully crafted to serve simultaneously as scientific and commercial objects. This meant appropriating existing notions of fidelity and realism and harmonizing them with scientific notions of objectivity or authenticity. It also involved the reconstruction of the field site itself into a place of scientific and technological control in what I have termed a "sterilization" of the soundscape of the field. Such sterilization enabled recordists to mediate strategically between the acoustic and epistemic properties of the field, the laboratory, and the sound studio.

In chapter 4, I show that the reconfiguration of these recordings into scientific objects involved an intricate social process. From the late 1940s onward, magnetic tape recording provided an alternative field technology that was less expensive and less complex to handle than the gramophone recorders. This granted a new mobility not only to field recordists, but indirectly also to their recordings: as recording became cheaper and easier, academic ornithologists collected extensive libraries of sampled natural sound, relying on a growing group of recordists. Examining the establishment and growth of one such sound archive, the Library of Natural Sounds at the Cornell Laboratory of Ornithology, I show that building up a stable stream of recorded sounds and transforming them into reliable specimens for scientific reference required considerable social negotiation. Based on existing work in the history and sociology of collaboration within science and between science and industry, I argue that the success of Cornell bioacousticians in aligning birdwatchers, amateur recordists, broadcasters, and diversely motivated biologists was grounded in a recognition of the wide range of uses that recordings could be put to by these respective groups. When collecting and exchanging recordings, ornithologists purposefully traded on their economic, social, and symbolic currencies. Drawing out the particular moral economy at play in the Cornell Library of Natural Sounds in the 1950s and 1960s, this chapter shows how in strategically trading recordings across various regimes of ownership, status, and access, scientists managed to reify a normative frame and their collaborators' endorsement of their values of mutual obligation and responsibility.

Chapter 5 focuses on the subsequent analysis of sound recordings. While magnetic tape instruments had opened up sound recording to a heterogeneous group of fieldworkers who contributed their recordings to dedicated sound archives, sound analysis itself became increasingly dissociated from the field site. By the early 1950s, academic biologists had adopted the sound spectrograph, a Bell Laboratories device that visualized short sound fragments, as their standard instrument of choice to investigate animal vocalizations "objectively." Sound spectrography performed sound analysis as part of a mechanical, visual, and expert authority that allowed additional control over acoustic phenomena. Yet meaningful interpretation of the visual language of sound spectrograms was also informed by the sonic skills and aural experiences of recordists and analysts. Trained judgment— visual as well as aural—thus supplemented the automatic registration of sound, but the trained ear served only to facilitate spectrographic analysis, not to replace it. Looking in detail at scientific imaging and visualization strategies, chapter 5 explores the strategies that biologists developed for converting hearing into reading and image into sound.

Chapter 6, finally, concludes with a synthesizing discussion that cuts through these chapters and again takes up the theme of mediated and unmediated listening as a mode of authoritative scientific observation. This chapter and the book as a whole open the history of scientific sound recording to attend to shifts and continuities that propel much of the history of science and technology, in relations between embodied listening and mediated recording, between amateur and professional scientists, between experts and laypeople, the field and the laboratory, the acoustic and the musical, and the auditory and the visual.

2   Scientific Scores and Musical Ears: Sound Diagrams in Field Recording

Musical Scintillations

> It is a mad, reckless song-fantasia, an outbreak of pent-up, irrepressible glee. He begins bravely enough with a number of well-sustained tones, but presently he accelerates his time, loses track of his motive, and goes to pieces in a burst of musical scintillations. (Mathews 1904, 49)

When writing this in 1904, naturalist Ferdinand S. Mathews was not reviewing an artistic composition. In fact, he was describing a motif of "musical fireworks" he heard produced by a bobolink, an American songbird. This was music, he explained, and in character easily comparable to some of Chopin's musical fantasias. In his *Field Book of Wild Birds and Their Music* (1904), Mathews compiled detailed song transcripts along with descriptions of the nesting and range of almost 130 wild bird species native to the eastern United States. The author of previous, lavishly watercolored field books on American shrubs and wildflowers, he had transcribed all birdsongs by ear and rendered them in a conventional musical notation. Mathews was by no means the first to write of birdsongs in such exalted terms—bird "music" was a trope of much naturalist writing before him.[1] Nor was he the first to place them on a musical staff.[2] Unlike his predecessors, however, Mathews emphasized the *scientific* nature of his endeavor. Anticipating that critics would doubt the value of recording birdsongs in musical notation, he first dismissed the conventional representation in nonsensical, often onomatopoeic, syllables, and presented his own glossary and introduction in an effort to mollify the skepticism of those unable to read music easily.

In a twist of irony, the very parts of the bobolink song that had inspired Mathews's powerful prose proved resistant to the musical approach he advocated: "I have never been able to 'sort out' the tones as they passed at this

Figure 2.1
Musical transcript of a bobolink song by Ferdinand S. Mathews.
Source: F. S. Mathews, *Field Book of Wild Birds and Their Music: A Description of the Character and Music of Birds* (New York: Putnam, 1904), 51.

break-neck speed," he explained, and "the difficulty in either describing or putting upon paper such music is insurmountable" (Mathews 1904, 49). Sketching the wildly bouncing notes to the best of his ability, he ultimately resorted to the transcription pictured in figure 2.1; its notes initially follow a traditional grid of quarter and eighth notes, but then oscillate quickly beyond the conventional dimensions of relative time and pitch, until they eventually recede—strangely—back in time. In all its unconventionality, his score might be read as an unsuspected precursor of postwar avant-garde notational experiments.[3] But however evocative Mathews's notation may appear to the present-day reader, to some of his contemporaries it suggested the inadequacy of musical notation for the accurate and intelligible representation of natural sounds.

Mathews's stretching of the syntactic capacities of musical notation in fact marks the beginning of a particularly lively intellectual exchange during which ornithologists widely debated the problem of notation and listening as methods of scientific study. At the turn of the twentieth century, interest in field observation of birds surged. Scientific ornithology long centered on armchair taxonomic and systematic study of stuffed specimens in museum collections. The study of living birds' habits, in contrast, had become the province of amateur naturalists—natural historians in the civic realm of schoolteachers, civil servants, writers, and pastors—along with a growing group of bourgeois birdwatchers.[4] That division blurred in the first decades of the new century, when, under influence of a new generation of professional ornithologists, fieldworkers increasingly sought to grant their investigations the status of scientific study. But while birds' migration patterns, nesting rituals, or food preferences seemed readily amenable to systematic scientific description, the specific qualities of birds' vocal repertoires proved more elusive and difficult to convey. Naturalists had traditionally been content with verbal descriptions and syllabic or onomatopoeic renditions, but between 1900 and 1930, field observers came to reconsider such methods and began instead to experiment with highly innovative musical transcriptions and graphic diagrams.

This methodological tangle gave rise to a debate regarding the variable boundaries of an emerging community of field listeners. As amateur and professional ornithologists sought to standardize a technology for representing natural sounds and to calibrate listening practices, they also came to define what it meant to listen to birdsong and to record it (that is, transform it into script) *as scientists*. Who should be allowed to listen with authority, and what kinds of professional literacy or perceptual skills could be expected of aspiring observers? A substantial strand of work in the history and sociology of science has shown that new forms of knowledge codification can be powerful factors in shaping and consolidating local scientific and technical communities.[5] At the same time, it has become clear that before such research tools can be usefully applied, various forms of nontextual transfer of skills take place, often within highly localized training regimes.[6] In the first decades of the twentieth century, therefore, ways of listening to birdsong took shape between two competing forces: efforts to standardize sound recording into a scientific methodology, and efforts to accommodate the multiplicity of interests and variable competence levels of listeners in local contexts. This chapter traces how notations were devised to straddle that tension.

Examining their role as "paper tools" in the development of field ornithology and scientific listening, it shows how notations allowed their users to listen in different ways and with different intents.[7] In the process of recording—that is, transcribing—on paper, these diagrammatic technologies order phenomena, by helping to eliminate "background" and electing certain features to foreground. That order is both epistemic and social. By their very syntax—"their visual form, rules of construction and maneuverability"—these schemata could thus be fitted to reify certain conceptions of bird vocalizations as, for instance, musical or expressive behavior, at the expense of alternative interpretations. But by their maneuverability and ease of application, such rules also help to negotiate a disciplinary authority or to construct a community of listeners. Understanding how listening and recording became a scientific technique in field biology, therefore, requires us to read these notations not just as inscriptions of birdsong, but as the outcome of a complex negotiation of conflicting demands for accessibility, accuracy, attractiveness, readability, and didactic potential.

The Attractions of the Field

In the half century between 1880 and 1930, the practices and interests of ornithologists shifted. Until the late nineteenth century, scientific ornithology had focused primarily on faunistic and taxonomic description and classification. Over the course of these fifty years, such interests—and the associated practices of shooting birds and collecting and conserving bird specimens, eggs, and skins for cabinet study—gradually began to give way to field studies and observations of behavior and ecology.[8]

In the historiography of ornithology, the shift has been framed as part of the professionalization of ornithology. That history has been traced by following disciplinary formation and conceptual transformations through the prism of several key protagonists.[9] Yet the negotiation of a new and decidedly professional identity for ornithology did not concern academic zoologists alone: it also impacted the realm of amateur naturalist study. Much ornithological fieldwork arose at the margins of professional academic science, in the sphere of local natural history societies and schoolteaching, and increasingly also of birdwatching. Historians of ornithology and conservation have interpreted this late nineteenth-century turn to the field as part of a profound change in attitudes toward nature.[10] Generally, the American and British middle classes began promoting a less exploitative and more harmonious relationship with nature. Urbanization and industrialization had made the presence of unexploited nature less self-evident than ever before, and the field study of live exemplars for pleasure seemed to provide an answer to a new yearning.[11] Changing leisure patterns for the middle class, too, had created a void to be filled with meaningful recreation, while the wider availability of bicycles and automobiles increased the mobility of new country dwellers and their ability to penetrate farther into the countryside. These changing attitudes found expression in the successive emergence of several bird protection movements in the last quarter of the nineteenth century (Allen 1976). In elite Victorian culture, birdwatching was co-opted to advance the cause of conservation and organized to form a respectable hobby.

The history of these protection movements and related changes in conceptions of nature go a long way toward explaining the attractions of the field for a large group of naturalists. As environmental historian Mark V. Barrow Jr. points out in his history of American scientific ornithology, these changing attitudes toward nature put an increasing strain on the traditional division of labor between professional specialists and a growing community of diversely motivated amateur bird enthusiasts (Barrow

1998, 17). By the early 1880s, scientific ornithologists had begun to foster a professional disciplinary identity of their own by clustering various local bird clubs into specialized national societies—such as the Deutsche Ornithologen-Gesellschaft, the British Ornithologists' Union, and the American Ornithologists' Union—and by organizing annual meetings and publishing periodicals.[12] In that process, scientific ornithologists were also beginning to differentiate themselves from the motley array of teachers, foresters, clergymen, and schoolboys who collected skins and eggs in their spare time. In principle, the societies' journals were open to any kind of contribution, but in the United States, for instance, the AOU's restrictive membership policy and its focus on technical discussions of nomenclature or faunistic description in practice kept amateur contributors at bay (Barrow 1998, 67–73).

In the 1880s and 1890s, the emerging bird protection movement achieved stricter regulation of bird shooting. The movement identified common practices such as hunting and commercial millinery as the main culprits for the rapid decline and possible extinction of several populations, but, importantly, they also blamed the collecting practices of professional ornithologists and their counterparts in the civic realm (Barrow 1998; Moss 2004; Orr 1992). In fact, ornithological societies such as the AOU themselves advocated stricter permit policies to protect their objects of study from careless shooting, but to protect their own collecting privileges they tried to exempt "proper scientific work" from more restrictive legislation—effectively denying the amateur birdwatcher, collector, and dealer a role in professional ornithological practice (Barrow 1998, 134–140; Davis 1994, 8–11). Faced with declining opportunities to collect and study actual specimens, naturalists instead published their observations of species they encountered in the wild (Barrow 1998, 172–175; Burkhardt 2005). These records comprised extensive descriptions of breeding habits, migration patterns, and mating or singing behavior, and even discussions of their implications for problems in Darwinian evolution (Barrow 1998). In the last decades of the nineteenth century, field ornithology and technical (taxonomic) contributions had been equally well represented in American ornithological publications, but in the first decades of the twentieth century, field reports began to dominate significantly (Battalio 1998, 94–110). In Britain, too, the fascination of birdwatchers shifted to detailed studies of the habits and life histories of common birds (Burkhardt 2005).

The growing confidence with which such field studies were carried out is exemplified by the *Wilson Bulletin* editor's call to readers to "prove that the slur often aimed at amateur field work is not applicable in your case at least. Such work needs to be done" (Jones 1905, 22). In fact, that "slur" persisted well into the 1920s, premised on a distinction reiterated by the author of a specimen identification tome, Robert Ridgway, in 1901: "Popular ornithology is the more entertaining, with its savor of the wildwood, green fields, the riverside and seashore, birdsongs and the many fascinating things connected with out-of-door-Nature," whereas "systematic ornithology, being a component part of biology—the science of life—is the more instructive and therefore more important."[13] Although professional ornithologists and amateur naturalists interacted more regularly than their counterparts in any other scientific discipline—the ornithological journal *The Auk*, for instance, contained a lengthy section of general notes where amateur ornithologists sent in their observations—the value of sight records and observations of living animals remained contested among professionals, especially when they had been recorded by amateur naturalists. For conservative professionals, unverifiable and often dubious identifications by inexperienced and potentially overzealous birdwatchers threatened to undercut the authority that their discipline had so carefully established. Some ornithologists still refused to accept sight records, especially those not substantiated by a preserved specimen, until the late 1920s. By that point, however, graduate training in ornithology at Cornell University and other institutions had begun to produce a new generation of professionals. These differed crucially from their technical predecessors of the 1890s, because they worked not only in museums but also in positions at universities and in state-organized biological surveys. Subject to different institutional and methodological demands than the typical nineteenth-century ornithologist, and with almost all species already described and mapped out, their methodological interests began to lean toward questions of ecology, population, and behavior, all of which required research on living birds.

As a result, they were increasingly willing to rely on their own field identifications and to accept observational records by others. In Britain, pioneering students of animal behavior such as amateur birdwatcher Frederick Kirkman and professional biologist Julian Huxley had begun to outline the contributions that field naturalists and birdwatchers could make toward the study of behavior, in gestures, singing, nest building, or migration, to resolve fundamental problems in biology (Burkhardt 2005, 98–126). What amateur naturalists needed, in their view, was knowledge of what to search for and a method to guide that search. Field observation, in other words, was a skill that could be learned. This point, that ornithology could profit from observations by field naturalists if only they were instructed correctly, was endorsed perhaps most explicitly by Ludlow Griscom. A graduate of one of the first ornithology programs, Griscom worked as the assistant curator of birds at the American Natural History Museum. At Cornell University, he had studied with the first professor of ornithology in the United States, Arthur A. Allen, and befriended the premier bird painter of the day, Luis Agassiz. Both were strongly committed to the popularization of their field. Although he was professionally concerned with systematics and bird distributions, Griscom—nicknamed the "Dean of the Birdwatchers"—repeatedly tried to strengthen relations between academic ornithology and hobbyist birdwatching (Davis 1994, 131–142).

He campaigned for the acceptance of field identifications by professional biologists, and his field guide *Birds of the New York City Region* (1923) introduced a new, more holistic method of attaining such identifications. Whereas the approach promoted in early field guides had been grafted onto the existing techniques of systematists, sorting primarily through color and visual pattern, Griscom instead advocated integrating knowledge of all the available information, such as locality, season, habitat, field marks, and also voice (Dunlap 2011, 72–82). To make their sightings and observations worthwhile, argued Griscom (1922, 39), students of field behavior required knowledge of such traits and above all a careful, scientific attitude: "If the bird student really wishes to make observations of scientific value, he must needs become a trained field ornithologist." But "to attain these qualifications calls for no special gifts or capabilities," and "granted no physical defects and some aptitude for the study, this is well within the reach of anyone" (Griscom 1922, 39–40). Based on his own field experience, Griscom concluded that only a very small portion of the apparent difficulties or inaccuracies of identification stemmed from what he termed "the human equation" and the range of possible variation in skill and aptitude among observers.[14] Such problems included defects of the eye and ear, as well as a lack of training. But they were greatly outnumbered by those caused by a persistent lack of attention: care, dedication, and a scientific attitude mattered more than virtuosity.

By the 1910s, a growing group of amateur naturalists, birdwatchers, and professional ornithologists had become emancipated as students of living birds in the field, relying on visual and auditory observations. But whereas taxonomic study had long provided a recognized protocol for the validation of scientific claims, through systematic comparison with carefully preserved and annotated specimens, observational records remained in a gray zone as no perceptual standards existed yet by which visual and auditory observation could authoritatively substantiate scientific claims. Existing historical scholarship has extensively addressed the role of field guides, such as those by Griscom and others, in shaping new standards and methodologies for the visual recognition of birds.[15] However, if ornithologists were to rely on their ears to locate, identify, and analyze bird behavior, listening too had to be learned. Just like problematic identifications based on sight alone, listeners' scientific legitimacy would hinge on their success in adopting a shared repertoire of terminology and techniques for listening and recording what they heard.

Avian Melodies between Science and Sentiment

By the early 1900s, such a repertoire had already been in the making. Against the shifts into a new era in conservation history described above, birdsong had assumed powerful currency among the growing group of naturalists, conservationists, and civic bird enthusiasts that had taken to fieldwork. In contrast to the visual attractions of birds' plumage and form, on display in scientific museums or as items of fashion, the melodious charm of their songs was celebrated as evidence of their beauty and an index of the pleasure and satisfaction that might be gleaned from studying living animals in their natural surroundings. In 1896 already, the American naturalist writer Charles Abbott (1896, 14–15) was glad to observe that "there is happily a wide-spread impression that birds are something more than mere 'specimens'. ... The woods, fields, mountain-sides and river-valleys tell another and a charming story." The experience of hearing "the music of the fields and woodlands" was extolled by the British ornithologist Charles Dixon (1897, i) a year later as "one of the most gratifying pleasures of the country. The variety of these songs is great, their beauty a refreshing and perennial one. [They] attract the least sentimental among us, arouse our sympathies, and charm the majority of us to a degree unapproached by any other living form." In fact, study of birdlife might not bring any material profit, the German teacher and amateur naturalist Ulrich Ramseyer (1908, 1) conceded, but "when I walk through field and forest, I am never alone. A little bird sings to me with inimitable joy its lovely luck."[16]

These authors' accounts of joyful encounters with singing birds during walks in the countryside tapped into a register that had become highly popular by the turn of the twentieth century. By the 1890s, the popular longing for a more vivid experience of nature was met by a growing market for nature essays, field guides, and bird books that encouraged readers to experience natural life firsthand. Their literary form often combined accurate naturalistic descriptions of animal life with a poetic style that centered on the authors' own experience (Kohler 2006, 73–76, 82–86). For many naturalists and their audiences, close observation of nature, and birdsong in particular, promised a brief glimpse into the deep structures and meanings of nature, and at the very least an aesthetic appreciation. But while personal and aesthetic revelation was central to nature writing of the day, it was not an invitation to sentimentality and artifice. This is evidenced among other things by a literary controversy that erupted in 1903, when naturalist writer John Burroughs chastised the authors of a popular genre of realistic animal fiction for producing "sham natural history" and for casting animal life in anthropomorphic and moralistic terms (Burroughs 1903, 298; see Lutts 1990, 40). The controversy was widely publicized and invited even a final word by the well-known outdoorsman Theodore Roosevelt, who famously denounced the contingent as "nature fakers" (Lutts 1990, 101). At stake was the reliability of nature writing; when the debate subsided around 1910, it had exposed the exploitation of nature description for sentimental purposes as deeply suspicious.

Against these broader literary discussions, much (popular) naturalist description took shape in what Robert Kohler (2006, 87) calls a symbiotic relationship between science and art. Around the turn of the twentieth century, then, "true" natural history writing purported to be both art and science: to provide emotional satisfaction through detailed observation and for scientific knowledge to enhance the aesthetic experience of nature (Kohler 2006, 74). The topic of birdsong and its recording thrived particularly well in this context. This is evident, first of all, from the surge in ornithological field guides such as Mathews's *Field Book* that concerned themselves with birdsong. Operating in the same cultural and textual domain of naturalist writing, these field guides fed the appetite of a growing readership of nature-goers, who were hungry for guidance on what there was to see, experience, and hear in nature; they provided detailed description and promoted outdoor observation. And although their authors were keen to underline their recordings of birdsong as the result of exacting observation, they did not shy from romantic or aesthetic overtones. Imaginative descriptions, musical notations, and onomatopoeic renderings all helped to convey not just a detailed understanding of specific birdsongs, but also their "compositional" intricacies and aesthetic nature. Repeatedly, Mathews criticized the birds he discussed for not taking their motif further to a logical end from a compositional standpoint or juxtaposed their songs with references to contemporary nursery melodies or Beethoven's Pastoral Symphony. Musical notation

seemed to authenticate the anthropomorphization of nature. As we will
see, for Mathews and many other naturalists of his time, it also provided a
means of recording and discussing birdsong in a way that expressed both
scientific and aesthetic aspirations.

Around 1910, however, birdsong did not just concern naturalist writers
or aspiring field ornithologists—it also resonated in a wide-ranging cultural
domain of music, pedagogy, and popular entertainment. Although across
these settings, birdsong had primarily been of interest for its popular and
aesthetic allure, here too, its practitioners recognized the symbiotic value
of accuracy and reliability as they shared with their naturalist counterparts
an approach to careful and detailed observation. For them, naturalist study
made the art authentic.

This was the case, first of all, in musical composition, where birdsong
was invoked to provide practicing composers with a rich set of materials. Amy Beach, for instance, one of the leading female composers in the
United States at the turn of the century, drew extensively on birdsongs.[17]
Beach, in fact, transcribed birdsongs in musical notation herself, as she had
been doing from an early age: as a young girl, she had been recruited by the
poet and English professor Edgar Rowland Sill on account of her talent for
absolute pitch and tonal memory. He took her to accompany him on a trip
around Berkeley to collect birdsong notations for a friend at the University
of California, who was writing a book on regional birdsongs. On a single
day, they took down the melodies of twenty birds with pencil and paper
(Tick and Beaudoin 2008, 326). In 1911, she was featured in a magazine
essay titled "Bird Songs Noted in the Woods and Fields by Mrs. H. H. A.
Beach for This Article," in which Beach—whom the author described as
having an extensive collection of bird melodies—was quoted describing the
sounds of several common birds, such as the wood pewee, song sparrow,
and thrush.

Such work formed the basis for several of Beach's compositions, including two well-known pieces in 1921, "A Hermit Thrush at Eve" and "A Hermit Thrush at Morn." She had transcribed the hermit thrush songs in her
New Hampshire studio: "In the deep woods near by, the Hermit Thrushes
sang all day long, so close to me that I could notate their songs and even
amuse myself by imitating them on the piano and having them answer.
The songs were so lovely and so consonant with our scales that I could
weave them into piano pieces" (quoted in Woodward 1923, ix). The calls in
these pieces, she insisted in a footnote, were exact notations, in the original
keys but an octave lower. In fact, as musicologist Denise von Glahn points

out, while Beach's work is often taken to express a "religio-romantic ide-
ology," the importance she attaches to the precision, rigor, and accuracy
in these transcriptions would align her equally with scientifically inclined
recordists of birdsong such as Mathews (von Glahn 2013, 43–44).[18]

Musical transcriptions of birdsong were not just aesthetically pleasing—
they were also utilized for pedagogical purposes, as is illustrated by William
B. Olds's *Twenty-Five Bird Songs for Children* (1914). Olds authored a series of
songbooks based on natural history subjects, which, as he outlined in *Bird
Songs, Flower Songs* in 1916, were meant specifically as a didactic resource.
Building on recent developments in music teaching, which encouraged
children to learn musical scales by associating each tone with a particular
color, Olds set out to connect tones with *actual* birds of a similar color,
whose stylized songs moreover corresponded to a particular tone.[19] Its benefits, he insisted, would be threefold: first, training of the ear to recognize
tone relationships; second, training of the eye to differentiate between colors; and third, the awakening of an interest in nature study and bird conservation. In his *Twenty-Five Bird Songs for Children*, Olds (1914, v) noted
that by rendering birdsongs in musical notation, "the actual bird-melodies
thus unconsciously absorbed should inevitably lead to a keener delight in
the singing of birds, and a better understanding of their songs. A further
result of this knowledge, I hope, will be the promotion of a deeper interest
in the whole subject of bird-life, and the need of its preservation."

This didactic premise bore the mark of the nature study movement,
which reached its zenith in America and Germany between 1900 and
1920.[20] As a popular educational movement combining facets of science,
aesthetic appreciation, conservation, and traditional pedagogy, nature
study sought to enhance children's individual learning through hands-on
instruction using concrete objects and representations. Its aim was to stim-
ulate an understanding of natural life in its own environmental context.
As one of the movement's spiritual leaders, Anna Botsford Comstock, laid
out its philosophy in her influential essay "The Teaching of Nature-Study,"
"Nature study aids both in discernment and in expression of things as they
are. [It] cultivates in the child a love of the beautiful; it brings to him early
a perception of color, form and music. He sees whatever there is in his environment. … Also, what there is of sound, he hears; he reads the music score
of the bird orchestra, separating each part and knowing which bird sings
it" (Comstock [1911] 1939, 1–2). For those seeking to cultivate in children
a power of accurate observation, birdsong provided a particularly appealing approach to instruct them in the "deep structures of nature." To be

pedagogically effective, however, notations were to be both appealing and as accurate as possible. Complementing his own transcriptions with notations by Mathews and Cheney, Olds (1914, v) had therefore transposed the melodies to keys more suitable for a child's voice, but insisted that the spirit and melodic intervals had been kept "absolutely true." An endorsement by Henry Oldys, a former assistant biologist with the Biological Survey of the Department of Agriculture and an expert on birdsong, further commended Olds's booklet for "preserving with exactness the bird themes he embodies in his songs" while harmonizing them so attractively (Olds 1914, iii). Just as Beach had for her musical scores, therefore, Olds had prepared his notations to be recognized both as an accurate transcription and as a helpful prescription for musical performance.

Henry Oldys was in fact perfectly situated to praise Olds's birdsong "transcriptions." He had been employed as a state ornithologist with the US biological survey for the Department of Agriculture, but later dedicated his full professional attention to lecturing and writing on the subject of bird music. While he continued to advertise his findings among ornithologists at professional meetings and in technical articles, he earned his living through a series of popular articles and well-attended popular lectures on bird protection and the musical nature of birdsong, performing occasionally on the Chautauqua circuit. The Chautauqua movement had originally started as a summer assembly of Sunday school teachers near Chautauqua Lake in New York in 1874, involving teacher-training classes, musical entertainment, lectures, and recreation. Within a few years, it had become overwhelmingly popular, and by the turn of the twentieth century, the idea of adult education and uplift for a broad audience quickly changed into a commercial affair, with performers traveling the country (though mostly the rural Midwest) along established circuits. The standardized education and entertainment programs included public performances by teachers, preachers, politicians, musicians, orchestras, circus artists, magicians, and artistic whistlers. The popularity of the circuits peaked during the 1910s and 1920s; their demise was heralded by the Great Depression and the rise of the film industry in the late 1920s (Canning 2005; Tapia 1997).[21]

Oldys had personally collected notations of hundreds of songs, which he reproduced during his lectures, interspersed by humorous anecdotes and following "the exact notes uttered by the bird" as nearly as possible (Oldys 1904a, 2). Listeners reportedly delighted in his marvelously perfect imitations, even though Oldys billed himself as an interpreter, not an imitator of bird music.[22] The audience could hardly be blamed for missing this subtle nuance, though. Whistled bird imitations, after all, were a fixture of popular entertainment. As Jacob Smith (2015) shows, the whistling genre had its roots on the vaudeville stage, but in the 1910s it was experiencing its golden age. Bird mimics added luster to educational initiatives and events sponsored by the bird protection movement, self-proclaimed artistic whistlers began performing in concert halls, and recordings of their acts proliferated in early phonograph catalogs.[23] Like Oldys, many such whistlers claimed to have based their act on extensive study and interaction with birds, and like his whistled interpretations, their performance often wedded popular entertainment with an ethical stance on conservation.

The genre was so popular, in fact, that in 1909 Agnes Woodward established the California School of Artistic Whistling, where (mostly female) students hoped to master the art of musical whistling. Although Woodward and her followers were careful to distinguish the art of whistling from plain, nonmusical bird mimicry, her teaching method too was grafted strongly on careful study of singing birds: like Beach, Woodward claimed to have developed her method for imitating birdcalls by spending time with the birds in a California hill cabin. To annotate and learn birdcalls, she had developed her own elaborate notation system; regular musical scores were annotated with syllables and a set of graphic symbols that sought to integrate the peculiarities of vocalizing birds into a prescriptive score for musical performance using the performer's own voice. Outlining this notation and her method for teaching artistic whistling to the general public in her 1923 book *Whistling as an Art*, moreover, Woodward insisted that whistling was actually not just an art or a vocation—although many of her students would consider it such—but also an educational practice that developed in her students the power of observation and imitation, and would lead one to a closer study of birdlife and their habits. In fact, the notations presented in *Whistling as an Art* explicitly connected the series of whistling exercises to various existing birdcalls, which allowed students after just a few lessons to imitate the elaborate calls of species such as the whippoorwill or indeed the song sparrow.

Tracing musical notations of birdsong along this continuum of musical composition, education, and popular entertainment illustrates both the ubiquitous interest that existed in birdsong, well beyond the confines of popular ornithological study, and the attraction of the musical score as a preferred technology for recording and conveying its intricacies. Music fostered a shared repertoire of recording, listening to, and thinking about birdsong in a way that tried to be at the same time aesthetic and naturalist, popular and scientific in nature. Musical notation sourced birdsong for

practical purposes, purposefully underlining its aesthetic and musical qualities. A record of birdsong could serve simultaneously as a didactic resource
for teaching musical affinity or imitative skill, or for prescribing musical
performance, just as it could be invoked to allure readers, drawing on its
musical and aesthetic associations to reinforce an interest in birdlife and
conservation ethics. And it was specifically the understanding that these
notations had been derived through careful study of actual bird behavior
that granted them a capacity to teach or allure. Musical notation therefore entailed an imitative, mimetic function—a way of recording birdsong
accurately and precisely. When naturalists, birdwatchers, and professional
ornithologists embraced musical notation to afford their aural observations
of birdsong a new kind of witnessing authority, they invoked its currency
as a medium for precise recording and its proven didactic potential. But in
drawing on musical notation as a way of rendering their observations more
scientific, they also came to share this textual and cultural domain with
composers, teachers, and popular entertainers.

Naturalists, Musicians, Scientists

Aspiring field observers had already begun to cast listening as an approach
for studying birds in the field, between the polar attributes of science and
sentiment. As early as 1879, naturalist Xenos Clark observed that birdsong
had "almost exclusively been treated of in the world of sentiment, where
poet-naturalists and nature-poets have culled a wealth of fancies." In contrast, he had attempted to compile notes on birdsong from the research
literature and from his own and fellow naturalists' observations to "make
inductions of scientific value" (Clark 1879, 209). In the early 1900s, American naturalist J. J. Williams (1902, 12) returned to this plight, complaining
that the common method of observing birdsongs "is to go into rhapsodies
over the enchantment of some bird's songs, the soul stirring melodies of
others, or the sad sorrowful intonations of others, exactly as we do with
human singers, … the writer securing his or her basis for such a treatise
from a week's visit to some neighbor's country home." The alternative, he
explained—listening to the songs' innumerable variations and functions—
was "more thorough, a little more intricate" and, by implication, more scientific. But this, warned Williams (1902, 12), was hard work: "To listen to
the ups and downs of a bird's song is easy for anyone to do, but to mentally
photograph all or any of these variations so that the mind can partially
recall them later on, is a task for even a practiced observer."

In the course of the 1910s, the contours of such a practiced, scientific observer became more precisely circumscribed. "Because of its difficulty," ornithologist and public speaker Henry Oldys (1916, 20) suggested in a piece in *The Auk*, the study of bird notes "should be undertaken only by trained musicians." Of course, he admitted, "much excellent work has been done by naturalists who lack musical training," but "the final word must be spoken by the musician, whose education fits him to observe important features that are quite certain to escape the attention of one whose musical ear has never been cultivated." His characterization of the ideal scientific observer as a trained musician echoed elsewhere, too. In Germany, the surgeon, leading entomologist, and amateur ornithologist Hans Stadler worked together with composer and reform educationist Cornel Schmitt to develop an "improved" musical notation, which they advertised in leading European ornithological journals as "a precise and scientific way of comparison" (Schmitt and Stadler 1913, 394). Similarly, Witmer Stone, the editor of the ornithological journal *The Auk* and a council member of the American Ornithologists' Union, blamed the failure of birdsong studies to advance along truly scientific lines on a lack of basic musical knowledge. In his view, musical notations constituted the "specimens for this line of investigation," and as such they were "absolutely essential, just as mathematics is essential in computing averages and percentages of error in bird migration, or chemical notation in recording the composition of pigments or other products of the bird's structure." Just like complex formulas, Stone explained, these notations might well be unintelligible to one who is ignorant of them, "but a knowledge of them is necessary to investigation" (Stone 1913, 473).

For self-declared musician-ornithologists like Stadler, Schmitt, and Stone, listening skill was not simply an extension of an acquired intimacy with species' sounds—it was a distinctively scientific technique. They presented auditory observation as structured by convention—not by subjective and artistic judgment, but by the systematic attitude required for precise notation, comparable to accepted practices of ornithological study. Like other scientific practices, musical listening entailed considerable technique. For that reason, field guides sometimes provided guidelines for naturalists on how to listen to and record bird vocalizations musically. Mathews (1904, xxi) included a musical key, tellingly subtitled "extremely important to those who do not read music," as the opening chapter of his *Field Book* (see plate 1). And a pamphlet that Schmitt and Stadler coauthored "for the recognition and investigation of birdsong" is probably one of the most explicit in its pedagogy. Its direct approach to addressing a lack of basic musical knowledge in the reader may be ascribed to Schmitt's staunch advocacy of the nature study movement in Germany (Neumann 2005). Having developed the method for transcribing birdsong, he later published several handbooks and field guides in which transcription was extensively discussed, while Stadler (who reputedly had perfect pitch) was especially keen to establish this transcription system as a scientific method (Schmitt 1923, 1932). The pamphlet proceeded first through two substantive chapters of introduction to prospective listeners before arriving at a guide to species songs: first through a step-by-step explanation of the basic principles of musical notation culminating in a quick quiz, and second through a discussion of the components of bird vocalizations, each part—voice, song, call, rhythm—closing with an elaborate list of attention points and hypotheses with which the aspiring observer could commence their research (Schmitt and Stadler 1919). These instructions provide the observer with a framework that structures both the reader's perception—"Now we want to attend to the accentuation! That is right, the second, deeper tone obtains the accent"—and attention (Schmitt and Stadler 1919, 6). They encouraged the naturalist to follow a specified routine of listening to, imitating, checking, and analyzing sounds, thus underlining the systematicity of musical listening.

Even with such basic introductions to musical syntax, it is doubtful that readers without prior musical knowledge will have picked up on its specifics. Moreover, even for those who, like Schmitt and Stadler, boasted that grasping these principles would take only a few hours, "reasonably good musical hearing" (1919, 1) remained an important precondition. For those who had that basis, however, musical listening could provide a standard lexicon and an interpretive frame that would enable them to coordinate and calibrate their aural experiences. Such systematicity, ornithologists like Schmitt and Stadler claimed, was what distinguished scientific from artistic listening. "The need of rigid accuracy and unbiased judgment must ever be kept in mind," Witmer Stone (1913, 474) exhorted his readers, to "guard against the enthusiasm of the musician which like that of the artist is sometimes inclined to run away with him when dealing with such problems." To that end, it was important to test musical records for their accuracy. Instead of being reconstructed from memory, valuable records "must be made by actual tests of each note with a graded pitch pipe, as is done by our best observers, while the time must be correctly gauged by some metronome contrivance" (Stone 1913, 474). The ideal of the scientist-musician made the field observer a methodical and detached listener—one who does not get involved in bird music emotionally or aesthetically as an artist would.

This reconfiguration of musical notation as an explicitly scientific technique for collecting and analyzing sound was an attempt to establish the scientific authority of aural field observations. Reference to a musically trained ear attributed to the field naturalist some form of competence in listening, suggesting what Thomas Porcello (2004, 734) has called "professional audition." The term denotes the auditory artifacts, techniques, discourses, and expertise that establish its possessors and users as professionally competent members of a community.[24] In this context, it alerts us to the ways such technical-musical discourse helped to establish a body of shared terminology for listeners to draw on, enabling them to describe or interpret birds' acoustic behavior efficiently and authoritatively in the field. At the same time, Porcello notes, such technical discourse may also have an exclusionary function for others who do not, or not yet, possess the knowledge or even the embodied experience implied by these codes. Just so, Oldys's and Stone's insistence on the importance and final authority of the trained musician in scientific discussions of birdsong *excluded* a large group of untrained listeners, whose ability to acquire a musical ear and sufficient competence for reading and writing musical notation, as we will see, was strongly contested. Nor was it clear exactly what musical listening as a scientific technique did *include*.

Interpreting Bird Music

Although its advocates promoted musical notation as a way to standardize birdsong notation, the musical score in fact gave its users a frame that was far less homogeneous than they acknowledged. Musical notation offered a versatile method. But differences in notation styles were not only formal, they also had far-reaching conceptual implications.

Although with musical notation a shared lexicon existed, it could not simply be applied "out of the box." Just as Agnes Woodward had embellished her musical scores with a host of self-invented symbols to represent bird whistles, ornithologists typically made modifications to accommodate the specific acoustic qualities of animal sound. Schmitt and Stadler, for instance, developed their own symbols for nonmusical sounds or typical song elements such as the canary's roller, added syllabic elements to represent timbre, and abolished the musical bar—thereby also abandoning the possibility of precise pitch notation (see figure 2.2). Others, like Robert Moore, adopted a more conventional approach to transcribing bird sound;

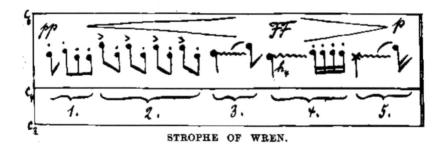

STROPHE OF WREN.

Figure 2.2
Adapted "scientific" musical notation by Hans Stadler and Cornel Schmitt.
Source: H. Stadler and C. Schmitt, "The Study of Bird-Notes," *British Birds* 8 (1) (1914):
7. Reproduced with kind permission of *British Birds*.

Figure 2.3
Musical transcription of a fox sparrow song by Robert T. Moore.
Source: R. T. Moore, "The Fox Sparrow as a Songster," *Auk* 30 (2) (1913): 178. Repro-
duced with kind permission of the American Ornithological Society.

they did fix the notes on a bar, thus suggesting that pitch could be fixed,
and added an absolute, metronomic measure of tempo, sometimes even
orchestrating the scores (see figure 2.3). German naturalist Alwin Voigt
(1913) applied basic musical notation for simple songs, but found that it
was completely unable to render more complex songs with small tone intervals. He thus complemented musical notation with a shorthand, Morse-like
script for short, long, and vibrating tones, highlighting rhythm but dismiss-
ing other parameters. In what was probably one of the most original and
dramatic revisions of musical notation, in the mid-1920s the British amateur naturalist Gladys Page-Wood developed a notation designed to achieve
the greatest possible accuracy. She eliminated conventional key, time, and
staff signatures and extended the five-line staff to represent microtones, on
the presumption that birds' musical scales were more detailed than those

of their human transcriber. By far her most original idea was to represent timbre through syllabic wordings and colored pitch lines, with various shades representing the nuances of sound quality in different bird voices: shades of gray or brown for tones that could be imitated by the human singing voice, and various shades of blue for whistled notes (see Hold 1970, 112).

In spite of the step-by-step instructions often provided in keys or introductory chapters, few of these proposed notations found application beyond the field notes of their original authors. New symbols were often burdensome for scholars to learn and for publishers to reproduce in print. Mutual understanding between field listeners was not only complicated by differences in the recordists' grasp of the fundamentals of music, but also by differences in the way they chose to present their recordings. Such tinkering with the tools of musical notation might reflect practical concerns (such as the need to make recording in the field less complex and burdensome) or accommodate differing methodological emphases (to highlight certain parameters over others). A further reason for the persistence of this variety in musical notations, however, was the existence of differing theoretical premises. Henry Oldys, for instance, used musical notation to demonstrate that the thrush and veerie songs he observed had rhythmical arrangements that were particularly pleasing to the human listener. He considered this proof that there was a universal appeal to musical appreciation among both birds and humans: "The bird expresses itself in human music. The notes were sung with great accuracy of intonation—my ear is very keen to detect variations from the true pitch" (Oldys 1913, 541). This could not be coincidence, ran Oldys's teleological argument: "Astonishing and revolutionary as it may seem, there is no escape from the conclusion that the evolution of bird music independently parallels the evolution of human music and that, therefore, such evolution in each case is not fortuitous, but tends inevitably toward a fixed ideal" (Oldys 1913, 541). In a comparable vein, German professor of ornithology Bernard Hoffmann adopted in *Kunst und Vogelgesang* (Art and Birdsong, 1908) what he called a "natural scientific-musical" approach.[25] By rendering the songs of talented individual birds in musical notation and syllables, he aimed to demonstrate that their song matched criteria for human music.

Oldys's and Hoffmann's work was by no means unique. On the contrary, it exemplifies a rich and multifaceted discourse on the evolution of music that had developed in the last quarter of the nineteenth century. This found initial support in Darwin's *Descent of Man* (1871). Darwin paid a great deal of attention to instances of nonhuman music, which he considered to have derived from the courtship display of birds. Their music, Darwin proposed, was not idle—it served a purpose of sexual selection of the fittest exemplars and ultimately the preservation of the species. By locating aesthetic sensibility in a fundamental mechanism of evolution, Darwin implied that birds and other animals (including humans) developed their disposition toward music early on the evolutionary scale, even before the development of language.[26] This position, however, was not universally shared. In Britain, Herbert Spencer had placed musical capacities last on his *scala natura*, a hierarchy in the advancement of mental capacities from animal to human: music had evolved from language and communication, and had progressed from simple tribal chants to complex Western polyphony. But such investigations into the evolution of animal and human music were not confined to biology alone; historians of music have recently shown that this discourse also had a broad appeal in turn-of-the-century comparative musicological and anthropological scholarship (for example, Ames 2003; Mundy, forthcoming; Rehding 2000; Zon 2007). Carl Stumpf's *Die Anfänge der Musik* (The Origins of Music) in 1911, for instance, revived the assumption that music evolved from signal calls.

Such musicological explorations of a common source for "primitive" folk music and "advanced" Western polyphony are in many ways analogous to naturalists' promotion of the study of birdsong as an exploration of the origins of human music. The British professor of zoology Walter Garstang (1922, 16–17), for example, concluded that "birds, aesthetically, are probably somewhere near the level of primitive man, and ... by the study of bird-song we may be enabled to retrace some of the steps by which the primitive emotional cries were transformed into the beginnings of artistic music." Pursuing this thesis, Garstang complemented his academic discussion on the mechanisms of evolution in birdsong with his own hand-drawn musical renditions and poetic interpretations. Some of these approaches may strike the contemporary reader as tautological, anthropomorphic, or speculative, but the same was not necessarily true of a turn-of-the-century naturalist audience. As Eileen Crist (1999) shows, naturalists since Darwin have assumed a conceptual continuity between the behavior of animals and of humans, often conceiving of this shared behavior as conscious or meaningful to the animal subject itself.[27] That only began to change profoundly in the late 1920s, when the pioneers of classical ethology, such as Konrad Lorenz and Nikolaas Tinbergen, developed a technical and objectifying idiom for observing and describing animal behavior that conceptualized it as compulsive, functional, and automatic (see Burkhardt 2005). This ethological perspective left little room for animal subjectivity and notions of "aesthetic consciousness." At least until the 1930s, however, the idea that musical sensibilities could be observed in animal behavior formed part of a broad spectrum of behavioral interpretations that seemed to be reconcilable with evolutionary thought.

This is illustrated by several exchanges in the ornithological literature. In the years after 1910, several prominent British naturalists quarreled over the precise behavioral role of birdsong, debating whether it was a factor in sexual selection, a mere instinctive ebullition of superfluous energy, or a matter of singing (as one ornithologist phrased it) "to please himself" (Kirkman 1910, 121). This comment provoked another ornithologist to declare that "it is surely going too far to grant aesthetic tastes to birds when the most generous of us cannot allow them in by far the greater number of our own species" (Stubbs 1910, 156). In the United States, comparable discussions on the evolution of birdsong unfolded, weighing the explanatory power of sexual selection against alternative explanations of imitation or simply enjoyment. Again, some ornithologists brought the factor of aesthetics into the equation: "How can we escape imputing the origin and development of this beauty in bird-song to an aesthetic sense in the birds themselves?" (Allen 1919, 531). Along similar lines, Richard Hunt, collector and assistant curator at the Museum of Vertebrate Zoology in California, advocated taking musical taste seriously as a factor in evolution. He had observed that birds seemed to want to "improve" their song, clearly preferring those sounds that the human observer had found "absolutely superior." Musical taste had a universal appeal. Hunt (1922, 196) concluded: "I believe that herein lies the explanation of the evolution of birdsong. The songster is an esthete."

If birdsong was inherently aesthetic according to Western musical standards, it was also to be represented as such. Students of birdsong could use musical notation simultaneously as a technique for recording songs in a technical and structured way and as an interpretive frame, assigning it a more literal role in their theories of behavior. Although the two uses often overlapped, they could also be applied independently of one another. Such versatility assured musical notation its currency in field studies of birdsong at the turn of the twentieth century. Yet the "professional jurisdiction" and explanatory power that self-declared scientist-musicians claimed were not unequivocally accepted by all their peers (Abbott 1998, 56). In the 1910s and 1920s, the application of musical notation for the recording of birdsong was received with growing reservations, both for its apparent lack of documentary precision in representing birdsong and for its manifest inaccessibility to people not trained in music or literate with musical notation.

Singing Out of Key

Criticism and polemics concerning the adequacy of musical notation to frame the study of birdsong found expression in several handbooks and journal contributions between 1900 and 1930, but it may be easiest to clarify the positions by zooming in on a brief but fierce controversy in 1915, when Aretas Saunders—a Yale graduate, former US forester, and beginning biology teacher—openly dismissed the scientific relevance of musical notation:[28]

> [Musical notation] has been made primarily for the recording and rendering of human music and birds do not usually sing according to such standards. ... Its standards of time do not allow the record of a song that does not follow the rhythmic beat of its measures. Do birds sing in any given key? Do they recognize any fundamental notes? Can one beat time to a bird's song? In the majority of cases these questions must be answered in the negative. ... The great majority of birds sing in a free, non-mechanical, natural manner that cannot be recorded on the musical scale with the exactness that it deserves. (Saunders 1915, 173–174)

Saunders was not new to music (he had learned to play various instruments at an early age), but his attempts to record birdsong on the traditional musical scales had been unsuccessful. To remedy this problem, he devised a new graphic method for recording birdsong, plotting values for pitch and duration on the coordinates of a vertical and horizontal axis. This, he believed, was "much simpler, and much more easily used and mastered" (Saunders 1915, 176) (figure 2.4 shows an example of his work). Like Stone, he had made a habit of using a small tuning fork or pitch pipe in the field to approximate the first note on the musical scale, and he also recommended using a stopwatch. Taken together, he believed, this would ensure both accuracy and simplicity. This does not mean Saunders dismissed everything musical. Though the absolute fidelity of musical notation was to be distrusted, he valued the skill of musical listening and the presence of a musical ear.

Saunders's initial ridicule of the musical method prompted a bitter public exchange of correspondence with Robert T. Moore. Moore, like Saunders, was an amateur associate of the American Ornithologists' Union. His primary engagement was with systematic ornithology, which became evident in an important collection of Mexican specimens he assembled. But

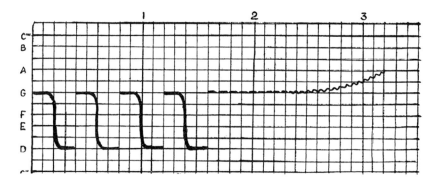

Figure 2.4
Graphic transcription of a field sparrow song by Aretas A. Saunders.
Source: A. A. Saunders, "Some Suggestions for Better Methods of Recording and Studying Birdsongs," *Auk* 32 (2) (1915): 175. Reproduced with kind permission of the American Ornithological Society.

he had recently been recording birdsongs too, using musical notation, and he took it on himself to defend this method against Saunders's allegations (see figure 2.3 for an example of his work).[29] Moore (1915, 538) criticized Saunders's proposal to replace a "splendid system of symbols, which has been evolved and improved by ages of use and is now better known to the public than any system of notation, used in the other departments of bird work," as simply a retrograde movement.

Moore's objections were not to a graphic method of representing birdsong as such, however—in fact, he insisted that musical notation was just as graphic as the alternative Saunders proposed. What he took issue with was the implication of Saunders's scheme: a reclassification of the features that ornithologists should aim to record in order to study birdsong. Saunders (1915, 535–538) had listed five features he deemed worthwhile recording: *pitch*, for which the unit of measurement was "of course the octave," divided into twelve half tones to show "their true relations"; *duration*, for which he proposed to record time in seconds; relative *intensity*, which in his method was represented by thickening the line; and *pronunciation*, represented by various wavy and zigzag lines (Saunders did concede that although tone *quality* or *timbre* was an important parameter, like most scores, his scheme did not record this dimension satisfactorily). Saunders's terminology contrasted with that employed by musical transcribers like Moore, who insisted that "pitch, time, intensity and quality" were the only worthwhile parameters. Moore (1915, 535) found fault with, among other things, Saunders's concept of duration, which obscured elements of time that were essential to the study of birdsong, such as meter and "the extremely important factor of 'rhythm.'" Saunders (1915, 177) had introduced the second as his measure, a "unit that is uniform and unchanging, and thoroughly understood both by musicians and by the uninitiated," to replace adagios, allegros, and metronome indications from which absolute duration first needed to be calculated. As a result, Moore (1916, 537) objected, the durations that Saunders had recorded had "as much value as the length of the white on the outer primary of a Junco. What we want to know about color is its arrangement or the relative proportion of the various colors on a bird."

Drawing on different idioms, the two men's exchange shuttled misunderstandings back and forth. In a last bid to convince Moore and the readers, Saunders (1916b, 103–105) wrote: "We must realize that it is our intention to study birdsongs, not from the standpoint of a musician but from that of a scientist, ... shall we change such a song in order to make it fit our method? Is such a proceeding scientific accuracy? Or is it the conception of a musician, so trained in the rules and necessities of human music that he is unable to conceive of music that is not rhythmic?"

Saunders's questioning of the documentary fidelity of musical notation paralleled a debate that was ongoing in the fields of comparative musicology and anthropology at the time. In the late nineteenth century, anthropologist Franz Boas had argued that ethnographic observers were unable to listen to the sounds of other cultural groups without filtering them through their own cultured set of perceptual biases (see Hochman 2010). By the turn of the twentieth century, ethnographers in the United States and Europe had increasingly come to acknowledge what they saw as a problem of perception and the incongruence between aural perception and its representation as written text. Ethnomusicologist Erich Moritz von Hornbostel, for instance, repeated Boas's concern in an 1911 article titled "Musikpsychologische Bemerkungen über Vogelgesang" (Music-Psychological Remarks on Birdsong), in which he invited his colleagues to consider the musicological interrelations between human and bird music. Musical notation, he wrote, risked "the most dangerous possibility of deception, to which even the most practiced musical observer falls prey time and again: intervals of the memory, that is, the familiar intervals of our music, which our ears add to [*hineinhören in*] the objectively given tone steps, even when these deviate considerably from them" (Hornbostel 1911, 119).[30] In response to this problem, Hornbostel (1911, 120) had turned to phonographic recordings, which "might, incidentally, also benefit the study of birdsong."[31] At the time Hornbostel was writing, actual phonographic reproductions of birdsongs in the field were still a relatively distant promise. Ethnologists, comparative musicologists, and zoologists began to incorporate phonographic recording techniques in the 1880s, but ornithological practice turned to phonographs only much later (Ames 2003; Brady 1999; Radick 2007; Shelemay 1991). This was mostly due to the limited mechanical possibilities of preelectric phonographs for amplifying the signals of wild (and mobile) animals. Birds could not be made to sit still as human subjects could. As chapter 3 will show, it was not until the 1930s that the phonographic ear began to surpass the musically skilled ears of Saunders's contemporaries. A new group of professional ornithologists and recording engineers pitted the gramophone, as an embodiment of faithful and objective reproduction, against the subjective, physically flawed perception of the field ornithologist.

In the 1910s and 1920s, however, such concerns with Western musical score's perceptual biases resonated with field ornithologists. But many were less radical in their departure from musical structures or traditional syllables. More than a decade after Saunders's comments, when amateur ornithologist Lucy Coffin (1928, 97, 99) realized that "the natural scale and rhythm of the bird is not the tempered scale of the piano nor the conventional rhythm of our written music," she wondered whether "perhaps a new system of musical notation may be necessary—possibly the Chinese, with its center 'four-square,' with four inter-notes or the Gregorian five-tone scale." Only then did Coffin suggest that her peers look into the possibilities of the electrified phonograph. In fact, even when naturalists and ornithologists considered the possibilities of phonographic recordings, musical notation remained their default reference point. When Ferdinand Mathews announced in 1904 that to take down the bobolink's song accurately, ornithologists would have to "wait for some interpreter with the sound-catching skill of a 'Blind Tom' and the phonograph combined" (Mathews 1904, 49), he did not expect his musical notation work to be replaced by the joint effort of an exceptional listener and a phonograph, but simply to have it facilitated by mechanical means.[32] Instead, these ornithologists anticipated that notation and the phonograph would simply complement each other: repeating and slowing down the phonographic record would enable listening to be more systematic and thus render their written notations more accurate.[33] Saunders himself, despite his obvious dissatisfaction with musical notation, does not seem to have considered the phonograph as a tool for ornithological study until 1929, and then only in the vaguest sense. That Saunders invested so little time in exploring the possibilities of the phonograph may seem surprising today. But his advocacy of graphic notation had been inspired by more than a need for greater documentary fidelity. It also represented ornithologists' attempts to make their recordings more comparable and accessible.

Comparing and Communicating Birdsongs

If graphic notation could replace a biased Western musical scheme, Saunders also identified greater comprehensiveness and simplicity as key to his new method. Graphic notation, he averred, "records the song simply and naturally, and so graphically that anyone can see its meaning *at a glance*" (Saunders 1916b; emphasis added). In contrast to musical notation—whose relation to a sound was clearly symbolic (the shape and position of a note on a staff determined its value)—the "simple" graphic line did not seem to require an extensive key or glossary. For Saunders and his contemporaries, this syntactic difference accorded distinct epistemic and didactic advantages.

First, if one could grasp the representation of a bird's song "at a glance," a different way of understanding the song became possible. Graphic notation had impressed Saunders by reflecting the variability of the songs he was studying—variability both among and within the songs produced by individual birds. He illustrated the benefits of his method with the results of a preliminary analysis of twenty-seven records of the field sparrow collected in this way (1922). Musical notations had typically been used to record individual songs—the ornithologist-musician would either cast individual song fragments as representative of a species (as in early field guides such as those by Mathews, Cheney, and others mentioned above) or approach the songs as highly localized performances that merited attention on their own (as in Oldys's spectacularly musical veerie songs). Rarely, however, did they present the songs comparatively, alongside those of other individuals of the species. Although field naturalists had long been occupied with tabulating variations in birds' singing behavior, they had focused almost exclusively on average song seasons when particular species would burst into song. Now, based on a synoptic reading of his graphic records, Saunders began to determine the ranges of variation for these birds' actual songs: their duration, their pitch, common characteristics, and the composition of long notes and short, repeated notes. Once a good number of records had been produced, he surmised, such features would enable conclusions to be drawn concerning "the" field sparrow song, but these records would also raise new questions about the conditions of variation in that song.

In the ensuing decades, Saunders developed an extensive portfolio of graphic records—by 1929, he had assembled nearly three hundred records for the field sparrow alone and at least five hundred fifty records for the song sparrow (Saunders 1924). His contemporaries were also putting graphic notation to similar uses. In 1924, for instance, John T. Nichols of the American Museum of Natural History processed almost two hundred recordings of song sparrow songs that had been made by a naturalist named William Wheeler. Wheeler and Nichols's explicit aim was to study variation. They declared musical notation to be "almost out of the question" as a means of capturing such variation, because it "does not clearly show the construction of the songs" (1924, 444). Instead, they explained, they had used a very simple graphic scheme consisting of dots, trills, and upward or downward lines to indicate slurs (see figure 2.5). Although the scheme largely disregarded many elements (time, quality) that ornithologists conventionally used to distinguish birdsongs, Wheeler and Nichols argued that a simple graphic form usefully allowed comparison of different songs, by bringing out discrete variations and irregularities in their typical "construction" (p. 446).[34] To advocates of graphic notation like Wheeler and Nichols, the lack of detail and the symbolic annotations typically provided by musical notation were precisely what enabled sounds to be grasped instantaneously and compared synoptically. This was a valuable

Figure 2.5
Graphic transcription of a song sparrow song by Wheeler and Nichols.
Source: Wheeler, W. C., and J. T. Nichols. 1924. The song of the song sparrow. *Auk* 41 (3): 446. Reproduced with kind permission of the American Ornithological Society.

advantage when sorting through hundreds of records to compare selected song "types."

Wheeler and Nichols regarded their graphic notation as particularly conducive to studying types and their variations, more so than to the intricacies of the virtuoso feathered performer. As paper tools, therefore, musical and graphic notations gave new shape to their objects of investigation. The syntax of the graphic line resided primarily in its visual form, an ascending, descending, continuous, discontinuous, or wavy line that captured the temporal development of pitch synoptically and thus arrested it in time. By reducing an acoustic event to the shape of a line, the intuitive symbolism of Saunders's graphic notation allowed a straightforward comparison of individual songs along these parameters. In contrast, the syntax of musical notation—equally symbolic, but more extensive and specifically defined—enabled sounds to be represented with a greater resolution of detail in the shadings of pitch, attack, or rhythmic inflection. Although this ensured a fine-grained detail, musical notations were less conducive to synoptic reading. Even for competent readers, representation of sound as a sequence of musical symbols required scores to be read *in time*, which complicated their synoptic comparison.

The representation of birdsong to be understood "at a glance" had other advantages. Saunders (1915, 183) felt that his graphic notation was not only less constraining and more accurate than musical notation, but also "intelligible to musicians, and a little less 'like Greek' to those whose knowledge of written music is slight." In fact, such worries about intelligibility were widespread among students of birdsong. In the same issue of *The Auk* where Saunders and Moore exchanged their last correspondence, naturalist Ewing Summers (1916, 79) complained that "but few people, one in a hundred or more, perhaps, are musicians far enough advanced to be able to perceive clearly what would be meant by some of the characters that would have to be employed, even when explained at length." Other ornithologists, amateur and professional, claimed to have developed effective teaching and recording methods that did not require any musical proficiency at all.[35] Many of these notations were effective tools to communicate and teach birdsongs to as wide an audience as possible, yet that advantage often had to be traded off against the accuracy that ornithologists expected of their recordings.

A notation devised by the Canadian amateur ornithologist William Rowan is a good example. Rowan, a gifted amateur musician who was just embarking on his career as an experimental zoologist, complained in a review in *British Birds* that near-accurate methods for recording birdsong such as musical notation had severe limitations. Even in the case of adapted musical notation such as that of the naturalist Stadler and the educationist Schmitt, who had claimed simplicity, Rowan (1925, 14) found that it was "confined entirely to musicians," and was "therefore ruled out for the layman." And since even "trained musicians" were often not able to judge an interval correctly, musical notation was of little use anyway. In fact, it appeared to him that "an *accurate* rendering of bird-calls, with the means at present at our disposal, is entirely impossible" (Rowan 1925, 15). His alternative, a shorthand script that combined traditional phonetic syllables and graphic lines to suggest accentuation and relative pitch, therefore surrendered accuracy completely in favor of intelligibility (figure 2.6): "A graphic indication of the kind suggested here, expressing nothing more than relative values, would be far more useful for field-work than a more accurate musical version that only a small percentage of observers could hope to employ. Its scientific value may be nil, but its practical value is very great" (Rowan 1925, 18). The script, in Rowan's words, had the advantages of "simplicity, plasticity and adaptability" and especially that "everyone can read it but write it as well" (Rowan 1925, 14, 16). With this emphasis on readability, Rowan abandoned a claim for the accuracy of recording— the point of his graphic description was that "a song once heard and put down can be recollected quite clearly years after" (Rowan 1925, 18). Rowan was part of a growing group of ornithologists who came to dismiss musical knowledge as a recording technology, finding it too complex, exclusive, and unintelligible to be of use for the practical purposes of identification and observation by a broader population of ornithologists, amateur naturalists, and birdwatchers. Like Saunders, Rowan relied on the graphic line to convey key features of a song "at a glance." The combination of phonetic terms and graphics, moreover, was flexible enough to adapt to different

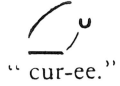

Figure 2.6
Graphic-syllabic rendering of a curlew call by William M. Rowan.
Source: W. M. Rowan, "A Practical Method of Recording Bird-Calls," *British Birds* 18 (1) (1925): 18. Reproduced with kind permission of *British Birds*.

kinds of sounds without having to annotate the score endlessly with new symbols and terms.

Like musical notation, then, these various versions of graphic notation (along with their syntactic constraints and affordances) were instrumentalized in a particular approach to studying and understanding birdsong— but simultaneously, they were also implicated in the configuration of a very heterogeneous group of aspiring field observers. Different notations presupposed different competencies and technical skills. As such, these paper tools not only mediated standards of accuracy and precision, but also questioned the claims to professional jurisdiction made by self-styled "scientist-musicians." Such denunciations of musical knowledge should not, of course, obscure the fact that listening remained an essential skill in the field observer's toolkit, regardless of the notation that was used. But while musical notation gradually disappeared from public awareness after 1930, for a long time disparate graphic, syllabic, or phonetic recording schemes remained the only instruments for field observers as they learned to study birds. The standard paper tool that ornithologists had sought in the first decades of the twentieth century did not come about until at least 1950, when the sound spectrograph seemed to provide a universal language for recording and analyzing bird vocalizations. However, as chapter 5 will show, even then some ornithologists remained deeply dissatisfied with the balance between objective analysis, efficiency, and intelligibility offered by the instrument.

Multiple Functions

Ornithologists at the turn of the century experimented with a host of different transcription techniques and combinations thereof—verbal description, musical notation, syllabics, onomatopoeia, phonetics, and graphic lines—and their proponents could not be neatly divided into opposing camps, each pursuing its own contained agenda. Rather, they covered a spectrum that, as laid out above, reached from the self-proclaimed scientist-musician's calculative and systematic method to the musical layman's accessibility and intuitive understanding. Caught between these different needs, fieldworkers struggled to devise a notation that could be both, at the same time. In accommodating such diverse interests and goals, these notations served not just as "inscriptive devices," but also functioned as "conscriptive devices."[36] If inscriptions may be understood as visual instantiations of knowledge that represent and package information through processes of refining, filtering, coding, comparing, and mathematizing in order to "harden" claims in a cascade of ever more refined representations, the notion of conscription highlights the other social functions that such inscriptions may be accorded in a social process. Thus understood, notations and diagrams serve as rhetorical objects that establish the credibility of their authors' claims, and equally as symbolic places that, in Wolff-Michael Roth's words, "bring together and engage collectivities to construct and interpret them." They provide "the material grounds over and about which sustained interactions occur, and which serve in part to coordinate these interactions" (Roth 2003, 18). Although the interactions facilitated by conscription devices have typically been observed between users in close proximity, the notations and diagrams described in this chapter also aimed to sustain interaction in a dispersed group of recordists. For their authors, they were often a means to calibrate, enact, and focus prevailing listening practices, to forms of implicit knowledge explicit, and to solicit the participation of individuals with diverse interests and intents. The notion of the conscription device thus alerts us to the multiplicity of functions that notations assumed in the practice of ornithological field recording.

There are numerous historical examples of scientific representations taking on different functions for different users. Based in part on this literature, the examples described in this chapter may be categorized as mnemonic, mimetic, didactic, or alluring (or as combinations of these categories). As David Kaiser (2005a) has shown, the Feynman diagram originated as a convenient mnemonic tool in physics, but acquired a new and initially unintended mimetic sense for a later generation of academic users.[37] Likewise, the transcriptions discussed above took on both mimetic and mnemonic roles. As mimetic records, in the minds of their users, notations implied a direct connection with the sound phenomena to which they referred and articulated a heightened sense of realism. This type of notation, especially musical scores, was predicated on the assertion that its writing conventions stood in as icons for the real thing, even though it could evidently do so to very different degrees: a musical lexicon could at the same time serve as the most accurate visual representation of a sound event and as concrete evidence of a musical sensibility in birds. As mnemonic devices, on the other hand, notations did not necessarily make such claims to realism. They emerged as part of an often highly individual and ad hoc scheme of perceptual signposts, and as such were not intended for analysis but for recognition, memorization, and self-instruction. Mathews's bobolink recording, for instance, would most likely have been unintelligible to those who did not have any prior notion of the actual sound, but could have considerable value for those who did.

Moreover, notations not only helped recordists orient themselves in the field, but also served to teach birdsong to others—lay readers—as in Rowan's graphic script. As didactic devices, such representations were often guided by similar demands for ease of use and intelligibility. But as an aid to printed instruction and the standardization of naturalists' auditory perception, they required a codification that was more fixed and rule-based than intuitive. Like mnemonic devices, however, didactically oriented recording systems depended on recognizability rather than detail. As Saunders explained, without the defining axes of time and pitch, his system would lose much of its precision but might still be used for teaching field students birds' acoustic signatures. Finally, as part of popular field guides as well as song or poetry books and educational handbooks, notations of birdsong also had a subtly alluring element for their various readerships. Anne Secord's work on nineteenth-century British botany is instructive in this regard (Secord 2002). She demonstrates that as botany turned to engage the amateur participant, expert naturalists began to see botanical plates no longer exclusively as a means to convey scientific truths to a specialist audience. The aesthetic and cultural appeal of the illustrations enabled them to be deployed as a way of teaching middle-class novices and recruiting them to extend their participation in scientific botany. Appealing to popular notions of pleasure and entertainment, in short, the notations' initial mimetic functions were complemented by didactic and alluring elements.

To appreciate how a community of practice might begin to gravitate to and cluster around ways of listening to birdsong in the field, then, we must attend to more than their scientific functionality alone. The diagrams and descriptions used by naturalists and professional ornithologists often followed similar visual schemata (such as musical notation or syllabic onomatopoeia) that rendered birdsong in similar forms across academic papers, birdwatchers' field guides, musical scores, poetic reinterpretations, or popularizing nature books. Their visual form and syntax placed these diagrams within a continuous series of scientific and popular records, allowing them to embody a variety of functions for very different audiences. It is this multiplicity that may help us to understand how a dispersed and heterogeneous community of listeners—of budding birdwatchers, amateur naturalists, pioneering ethologists, and museum ornithologists, all of different feathers—began to coalesce around birdsong. But that multiplicity also crippled the many attempts to establish a single, comprehensive diagrammatic form for recording birdsongs in the field. Demands of accessibility, flexibility, accuracy, readability, and didactic potential sometimes proved contradictory and were often difficult to integrate into a standardized and optically consistent scheme. Although from 1930 onward, electrical recording instruments would promise to solve such tensions, these tensions continued to shape how biologists and their collaborators engaged with birdsong.

3   Staging Sterile Sound: Producing and Reproducing Natural Field Recordings

Soundscapes and the Field

"There is little question that if it were possible to produce satisfactory phonograph records of birds' songs and calls, the study of bird voices would be greatly stimulated. All previous methods, while useful in their way, were, at best, merely makeshifts, awaiting the time when science should have advanced sufficiently so that faithful reproductions of actual singing birds could be made" (Brand 1932, 436). Thus Albert Brand, a part-time ornithologist at Cornell University, described his ambitions for his project ahead. The failure to devise a standard notation for the study of birdsong had raised ornithologists' expectations of new means of recording. But when electrical amplification turned those hopes into a practical reality, such "faithful reproduction" proved a formidable challenge, which changed the nature of recording in the field. Capturing birds' song behavior on records required, after all, the relation between such behavior and the field to be reordered in significant ways. This chapter traces how, in tackling the challenges of recording during the 1930s and early 1940s, ornithologists—in collaboration with sound engineers and recording professionals—achieved a new kind of control over the field, both as an acoustic space and as a site where natural and authentic behavior could be studied.

Electrical recording inserted the field site in the networks that instituted what Emily Thompson (2002, 1997) has called a "soundscape of modernity."[1] In the first decades of the twentieth century, scientists, engineers, and technicians developed a new language for describing the behavior of sound, new tools and techniques for measuring it, and new materials for designing and controlling it. In this age of electroacoustics, new technologies such as the microphone and loudspeaker, as well as sound-engineered environments, redefined what counted as signal and noise, and promoted a clear, controlled, and nonreverberant sound.[2] Dissociating acoustic phenomena from their spatial context further allowed the soundscapes of recording studios, auditoriums, and office spaces to be engineered with precision and care. But that soundscape of modernity extended well beyond the architecture of corporate and cultural America; it also stretched deep into the wilderness of field and forest.

This should perhaps come as no surprise, given that issues of isolation and control have often factored into the histories and designs of scientific spaces. Scientists have long sought to shield their experiments from acoustic and vibrational interference, by constructing sites far from the disturbances of urban life or by soundproofing their laboratories, a development that reached its apex in the anechoic chamber (Mody 2005; Schmidgen 2003, 2008). Noise—even acoustic noise—is potentially problematic for the production of scientific knowledge, because it ruptures the suggestion that the "ambient conditions" of one laboratory are similar to the next. Science and technology studies have related the wide distribution of scientific knowledge at least in part to the successful establishment of standardized contexts for making that knowledge, and the laboratory space serves as its prime example (Henke and Gieryn 2008; Shapin 1994). That is not to say that laboratories always *were* the same, but rather that their homogenization fashioned them into a singular and universal cultural space and fostered a belief in its "generic placelessness" (Kohler 2002b).[3] In maintaining an environment that is in all aspects, also sonically, sterile, laboratories uphold a presumption of equivalence: the conviction that places that look and sound alike will also perform alike.

And yet, that the outdoors should be subject to the same aspiration of acoustic homogenization is at least remarkable. In contrast to the laboratory's claim to placelessness, the scientific field site has often tended to draw its authority from the particular place it is situated in and bound to (Kohler 2002b). By bringing natural phenomena into the generic background of the laboratory, one hopes to divorce them from their original context, so that they can be examined, measured, and manipulated on the observer's own terms, independently of the unpredictable conditions of a situation "in the wild" (Knorr Cetina 1992, 1999). In the field, in contrast, this very particularity and variability are part of the object under investigation—it is what makes them valuable to scientists. The rise of the laboratory has rendered the field in its mirror image, as a place where scientists encounter their objects of study in the most natural and unadulterated state (Burkhardt 1999; Gieryn 2002; Rees 2005). But as a result of this rhetorical opposition, phenomena studied in the field have also often been considered much more multivariate, uncontrollable, and unrepeatable than those enculturated in the laboratory. When contrasted to the enclosed laboratory, as a space of knowledge production, the field site is permeated by undesired natural, social, cultural influences.[4] And as field biologists are joined by hunters, poachers, sportsmen, vacationers, or agriculturalists, not only do the discourse or epistemic virtues of the field change, so do its soundscapes. Indeed, recent work in environmental history, cultural geography, and the history of technology shows that the spaces that ornithologists adopted as field sites were intersected by different, often contested, moral geographies and their contradicting stakes in the sonic environment.[5]

To counter such influences of local ambience on their research, field scientists have sometimes sought to emulate the standing of laboratory science, by adopting its methods or by containing its spatial environment, often with mixed results. Just so, I argue, field recordists tried, with varying degrees of success, to filter, segregate, and control the complex acoustics of the sites they adopted as research environments. It is in this sense that the field site has undergone a process of "laboratorization," a term I use with reference to historian of science Robert Kohler (2002a). This and other studies of field science show that the laboratory and field site as described above are actually ideal types and part of a continuum with plenty of traffic and hybrid forms in between.[6] Others have identified, moreover, similar processes at work in making nature knowable in the field and the laboratory (Latour 1995; Roth and Bowen 1999). Just as laboratories could be "naturalized" or field sites became dependent on laboratory-like methods, I argue that electrical recording introduced a set of epistemic practices in the field that conventionally identify with the laboratory. These interventions aimed to establish a greater control over the field environment, akin to that permitted by the laboratory or the sound studio, which ultimately facilitated the transfer of sound recordings as objects of scientific study *into* the laboratory. This is not to say that the laboratory ideal supplanted that of the field site. In fact, fieldworkers remained convinced that truly natural sounds could not be contained within a controlled, studio-like environment. Rather, new field recording techniques allowed their users to claim the epistemic authority that comes with access to unadulterated, authentic nature while still maximizing control over its (acoustic) conditions that ultimately domesticates its subjects.

In conceptualizing this dynamic, I take a cue from Thomas Gieryn's (2006) analysis of fieldwork by Chicago School urban sociologists, who presented their working methods as experimental laboratory practice as well as observational field practice. This doubling allowed them to invoke the epistemic authority of either place when they deemed it strategically appropriate. Gieryn identifies three dimensions along which sociologists shuttled back and forth between the *ideal types* of laboratory and field site as truth spots. First, objects of study are *found* naturally at the field site, whereas in the laboratory they first have to be *made* and *crafted* into suitable objects of research. Second, at the field site, the objects and phenomena of research are intimately connected to and part of the *place* of research, whereas laboratory practice intentionally constructs environments and objects that are *generic*, which is how their universality is suggested. Finally, field and lab position the analysts in different ways vis-à-vis their objects of study. At the field site, researchers become *immersed*—their research object is everywhere around them—whereas the laboratory creates distance between the researcher and an object of study, allowing its manipulation in a mechanical and antiseptic way. I use these dimensions—ready-found vs. crafted, specific vs. generic context, and immersion vs. detachment—as a starting point for coming to terms with the ways nature recordists situated their work in the field and at the same time reproduced that field as a new phenomenal domain. As nature recordists began reducing place and natural context to a generic and mute backdrop, discursively as well as materially, such seemingly stable categories as field site, laboratory, and studio began to leak into each other.

Bird Records and Soundbooks

Cornell Ornithologists and the Movie Industry
In the spring of 1929, the motion picture company Fox-Case Movietone appealed to the expertise of Dr. Arthur A. Allen, then an eminent ornithological educator at Cornell University, to help them record birdsongs. The company's interest in animal vocalizations had been piqued when improved phonographic techniques enabled the synchronization of sound and film around 1926 (Kellogg 1955). In an attempt to collect authentic sound effects for the soundtracks of a new genre of "talking movies," Fox-Case Movietone had on several occasions tried to record wild birds, but they had failed miserably. Birds, they complained, would simply not sit still in front of the microphone. Allen, however, had specialized in bird photography, so surely, the engineers hoped, he would know how to bring birds close enough to the microphone to record them. Eager to show off their fieldmanship, Allen and Peter Kellogg, a graduate student of Allen's at the "Laboratory of Ornithology," an informal unit within Cornell's Department of Entomology, agreed to help the recording engineers.[7]

Meeting them at Ithaca's Stewart Park on a very early May morning, the ornithologists found themselves confronted with a large recording truck, equipped with state-of-the-art microphones and electrical amplification systems. Sounds were recorded using a Western Electric "sound camera," a device that photographed acoustic energy as variations in light exposure on the side of a film reel. The technique proved particularly useful for the movie industry, as sounds could be cut and pasted to fit the scene or transferred onto disc for further release. The collaboration proved a success. According to Cornell lore, once Allen had cleverly convinced the engineers of the importance of patience when dealing with wild animals, they managed to record the songs of three birds within the hour. After hearing the Movietone recordings several weeks later, Allen and Kellogg concluded that the technology offered great possibilities for their teaching and study of ornithology (Kellogg 1938).

The contacts between the Cornell ornithologists and the movie industry effectively revived a fascination with phonographic recording of bird voices that had already occupied the previous generation of ornithologists. As early as 1898, a record of a brown thrasher had been presented at the American Ornithologists' Union congress, which the audience welcomed as a "new and unique feature" that suggested "great possibilities to be looked for in the future" (Judd 1899; Sage 1899, 53). However, the technical obstacles hindering these possibilities from materializing were formidable. The acoustic phonograph's limited mechanical means for amplifying the signal required an animal to sit still and perform directly into the recording horn. For that reason, many published recordings involved whistled imitations of birdsong, which preelectric recording engineers found the easiest of all sounds to capture (Copeland and Boswall 1983, 73). In 1910, however, the German amateur scientist and birdkeeper Karl Reich collaborated with an engineer from the Deutsche Grammophon Gesellschaft to record the songs of his caged nightingales. Located in the Harz mountain region, Reich had been engaged in its lucrative caged-bird and singing canary industry, using instructor birds, serinettes, or bird organs to train domestic songbirds to produce their famous "rollers." Using a dummy, Reich had managed to train his nightingales and a few other species to perform in front of the phonograph (Birkhead 2003). The records were an immediate success, leading them to appear on Victor's prestigious Red Seal format and the cheaper Green Label by 1911 (Copeland, Boswall, and Petts 1988, 9). Seeking scientific recognition, the Grammophon Company presented the records during a break from sessions at the fifth International Ornithological Congress in Berlin, where the attending ornithologists praised them more for their enjoyment value than for their scientific potential (Birkhead and van Balen 2008; Schalow 1911, 39). Reich's gramophone recordings dominated the market for over a decade, until electrical recording by the mid-1920s opened new possibilities. In 1927, HMV released a record featuring professional cellist Beatrice Harrison in a musical duet with a nightingale, an event that had originally been staged for a live BBC broadcast from her Oxted garden in 1924 and that had attracted such public interest that the event was staged every spring until 1935. Later recordings featured sound effects collected at the London Zoological Gardens (Copeland, Boswall, and Petts 1988, 5). Until 1930, the entertainment industry had taken the lead in developing nature recording into a practice that was technically feasible and financially sustainable.

While the Cornell ornithologists' interest in bird recording too had initially been framed by the movie industry's demand for sound effects, their collaboration was short-lived. Just a few months after the first recordings were made, the financial crash of 1929 stymied the industry's enthusiasm for collecting authentic sound effects on costly and time-consuming field expeditions. With Movietone's withdrawal from the collaboration, the Cornell ornithologists' access to expensive equipment and specialized technical expertise dwindled. Their commitment to wildlife recording was kept afloat, however, by one of their part-time students, Albert R. Brand, a retired stock trader turned amateur ornithologist. As an ornithologist at the American Museum of Natural History (AMNH) in New York and a special student at the Cornell College of Agriculture, Brand had become part of the inner circle of field ornithologists around Arthur Allen. Having come to believe that "many of the secrets of avian life are hidden in an understanding of the meaning of the song," he offered to invest his time and considerable private funds in organizing a small-scale portable recording ensemble based on the sound-film camera Allen had seen in action. As we will see, once the custom-built equipment arrived from the Audio Research Laboratories, making the ensemble work in the unforgiving conditions of the field site proved a challenge. By 1932, however, together with colleagues in the School of Electrical Engineering, Brand and Kellogg had managed to repurpose the recording equipment to suit their needs and to demonstrate its uses at the American Ornithologists' Union. Their recordings, Brand (1932, 439) was convinced, would "become almost as indispensable to the ornithologist of the future, as the camera has become to the present generation, and as the gun was to earlier workers" for shooting specimens.[8]

But although Brand and his colleagues' ambitions for sound recording were clearly scientific, his business instinct, combined with the department's commitment to public outreach and teaching, led the recordists to look for other applications soon after. Recordings and moving images from the 1935 expedition cosponsored by the AMNH and the Cornell recordists to document endangered species such as the ivory-billed woodpecker, for instance, were used to enliven taxidermic displays at the Cornell University Museum (Eley 2013, 16). By the late 1930s, such phonographic playbacks of animal vocalizations and ambient sound tracks even became a regular fixture at the AMNH. In addition, in 1934 Brand published the first of a series of gramophone records from their growing collection of sound films. While catalogs by commercial publishers such as Victor Records listed recordings of captive bird specimens, accompanied by orchestral arrangements or imitations by professional whistlers, these recordings were pitched as an aid to learning and identifying wild birdsongs at home or in the classroom. By 1938, Brand had already published five gramophone records—some with a conventional publisher and one in collaboration with the American Foundation for the Blind. As we will see in chapter 4, these recordings' popular (and commercial) appeal would become key to the recordists' investment in field recording.

By 1940, however, the recording project was put on temporary hold by Brand's death after a long illness and by new activities prompted by the onset of the war. Kellogg, who had dedicated his doctorate to the recording project, took over the project, but was immediately engaged to apply his newly won skills in organizing a "radar school" for engineers where they could become familiar with the latest electronic technology (Kellogg 1962a). His assignment at Western Electric would help him form close and lasting relationships with people at Bell Laboratories, who would be influential in developing new instruments such as the sound spectrograph (Little 2003). While laboratory staff members and students joined the army, Allen tried to maintain a regular teaching schedule. In 1943, Kellogg, Allen, and Allen's son joined the staff of acoustician Carl Eyring (1946), who had been contracted by the US Armed Forces to investigate the acoustic conditions of jungle environments. Experience had taught that enemy forces mimicked natural sounds for signaling purposes on the battlefield and the ornithologists joined the team in the Panama jungle (in preparation for an expedition to the South Pacific that never took place) to record its ambient sounds and inventory the sounds of identifiable species.[9] Finding, moreover, that sound film deteriorated quickly under tropical conditions, and frustrated by their inability to develop the film on-site, the recordists decided to change to instantaneous disc recording.

Ludwig Koch and British Naturalists

Around the same time, recording of wild birdsongs took hold on the European continent, as the entertainment industry reached out to established ornithologists to record and capitalize on nature sounds. By 1928, the record company Lindström AG in Germany commissioned former concert singer and military intelligence officer Ludwig Koch to head its new department of culture. Charged with the task of publicizing the educational potential of Lindström's gramophones and to prevent HMV from dominating the market, Koch pitched a new publication genre: the soundbook, a multimedia educational book that included images and gramophone records. During Koch's five-year tenure at Lindström, thirteen soundbooks covered urban soundscapes and historical topics, but also natural history. Himself an amateur birdwatcher and interested in phonograph recording since his youth, Koch engaged prominent German naturalists to collaborate with him, enlisting, for instance, the director of the Berlin Zoo, Dr. Lutz Heck, to publish two soundbooks, *Schrei der Steppe: Tönende Bilder aus dem Ostafrikanischen Busch* (Calls of the Savannah: Sound Pictures from the East African Bush, 1933) and the double-sided *Der Wald erschallt* (The Forest Resounds, 1934). Adopting the tone of a nature documentary, these narrated texts guide readers past a sequence of scenes, pausing with images and the sound excerpts on record. In 1935, Koch also collaborated with Oscar Heinroth, whom Konrad Lorenz would count among the founding fathers of ethology on account of his work in ornithology. The resulting soundbook, *Gefiederte Meistersänger* (Feathered Master Singers), collected songs of twenty-five common German bird species, which he had been recording since 1927, on three double-sided discs (Heinroth and Koch 1935).[10] Elsewhere in Europe, too, animal voices became increasingly popular topics for soundbooks and gramophone recordings. In the mid-1930s, radio engineers began experimenting with location recording on behalf of public broadcasters, and having turned their attention to animals, several of them continued to publish their work as educational records. Danish recording engineer Carl Weismann, for example, distributed his recordings to schools via Dansk Stadsradiofonien before he began selling them commercially himself.[11] In Sweden, Swedish broadcaster Radiotjänst AB engaged amateur ornithologist and private detective Gunnar Lekander to record birdsongs, together with its recording engineer Sture Palmér. Making use of Telefunken disc cutters, Palmér would release around sixty-five 78 rpm records of recorded birdsongs in the following decades.[12]

The changing political climate in Germany, meanwhile, compelled the outspoken Koch to flee to London in 1935. Within months there he was approached by the BBC and introduced to key figures in Britain's "new naturalism" movement. From the mid-1920s onward, a small group of ornithologists and naturalists had begun to encourage ornithologists to abandon specimen and sighting collecting for detailed description of the living bird, its behavior, ecology, and population dynamics. This, they expected, would integrate ornithology into a broad-based discipline of professional biology.[13] But although the process was initiated by professionals, Koch's new acquaintances were keen to enlist amateur birdwatchers and found in Koch an associate who could help direct the enthusiasm of "a vast army of bird-lovers and bird-watchers" toward careful observation and organized survey (Huxley 1916, 142).

Among those new naturalists was Julian Huxley, then secretary of the London Zoological Society and a prolific science popularizer, who had just produced the critically acclaimed ethological film *The Life Story of the Gannet*.[14] Huxley introduced Koch to Harry Witherby, a seasoned ornithologist and publisher of natural history publications, and his protégé, conservationist and natural history author E. Max Nicholson, whose popularizing works introduced a new standard for bird study (Koch 1955, 36–37). Nicholson had also been instrumental in founding the British Trust for Ornithology—a national council promoting coordinated research by British birdwatchers. With Huxley, Nicholson campaigned to foster scientific literacy among the British public, attempting among other things to refashion the image of the British birdwatcher from a lone naturalist into an amateur citizen-scientist who formed part of a tightly organized network of observers (MacDonald 2002; Toogood 2011). With help of this network, Koch was publishing soundbooks again by 1936, this time in collaboration with Nicholson and the record company Parlophone (like Lindström, a subsidiary of Electric and Musical Industries [EMI]). The result was a soundbook collecting calls of fifteen species on two double-sided gramophone records (see plate 2), and another three discs featuring twenty-one species in the sequel *More Songs of Wild Birds* the year after (Nicholson and Koch 1936, 1937). Koch went on to bring out other sound publications, which he recorded with Huxley in the London Zoo and together with the Belgian Queen Mother at the Royal Park (Huxley and Koch 1938).

As in the United States, the advent of another world war complicated but, in time, also invigorated field recording. After a brief stint as an interned enemy alien, Koch began working for the British Broadcasting Corporation's European Service and from 1942 onward produced numerous items for its Home Service, including "sound pictures" for programs such as *Country Magazine* and illustrated talks for the *Children's Hour*. Such sonically evocative features became a regular fixture on the radio during and after the war. Until the mid-1930s, radio producers at the BBC had addressed topics in natural history, science, or outdoor life through lectern-based exposés, but now they gradually began to consider entertainment value as well. The acquisition of new mobile recording facilities in 1938 enabled them to experiment more systematically with on-site recordings in ways that were sonically more evocative than the previous talking voices. Moreover, with television services largely suspended during wartime, field recordings captured the attention of radio audiences more than anything. Construed as innocent, uplifting, yet heavily politicized as an expression of British natural and cultural heritage, the British Broadcasting Corporation found birds' songs exceptionally suited to soothe the airwaves during and immediately after the war.[15] The success of these wartime broadcasts led the BBC to make natural history subjects a regular fixture in its programming and to develop long-running popular broadcasts such as *The Naturalist* (1946) and *Birdsong of the Month* (1947). These would greatly amplify the demand for a steady supply of new field recordings on a wide range of (ornithological) subjects.

As these sections show, pioneering wildlife recordists learned their craft with technical and financial support by motion picture companies, record producers, and book publishers, leading them to recognize sound recording's applications above all as a technology of education and popularization.[16] Such applications situate these first attempts at recording the vocalizations of wild birds within a broader history of cinematic and commercial recording. Yet these formative conditions would also shape the opportunities offered by the new medium for the scientific study of wildlife sounds.

The Difficulty of Listening

Like the notations discussed in chapter 2, sound recordings assumed multiple functions. Besides dramatically affecting exhibition or documentary uses, the ornithologists and recordists anticipated that sound records would have a significant impact on their own scientific studies of birdsong.[17] Soundbook authors were keen to emphasize this versatility. In the introduction to Nicholson and Koch's *Songs of Wild Birds*, Julian Huxley suggested that "to both [bird lovers and amateur scientists] as well as to professional ornithologists, this book will be of great interest and value" (Nicholson and Koch 1936, xiii). Specifying such value to the professional ornithologist in

Animal Language a year later, he predicted that "students of animal behavior will be able to make experiments on the differences between related species, on the share of what is innate and what is learnt in determining reaction. … In any case, it is clear that an interesting field, with both practical and theoretical sides, is here opened up by the advance of technique" (Huxley and Koch 1938, 6). Similarly, noting that "classes of bird students would find birdsong records extremely useful," Albert Brand (1932, 436) argued that this recording work would also open "the way to a multitude of scientific experiments [and] an entirely new field to the ornithologist," as now "the serious student of ornithology can study song in a way that has been impossible heretofore." The recordings produced by Brand and Kellogg and Koch and Nicholson were written into the scientific study of birdsong in at least two ways, each confronting in its own way the problematic nature of listening.

In presenting sound recordings as an aural counterpart to the identification guide, these ornithologists established, first of all, a standardized aural discourse (specifically, a discourse of bird vocalizations) that could be shared by both casual birdwatchers and professional ornithologists. Thus, in *Songs of Wild Birds* Nicholson and Koch included an informative appendix on techniques of listening to bird vocalizations. Of course, they wrote, field observers must have "tolerably good hearing," but to find out how particular songs differed in duration, intervals, rate of delivery, pitch, and the notes of which they were composed, listeners could "quickly develop a surprisingly good ear," provided that they kept in practice (Nicholson and Koch 1936, 189–190). "Anyone of reasonable intelligence" would then be able to contribute to answering important questions in ornithology: if listeners studied and reported differences in the song periods or between regional, seasonal, and individual patterns of song variation, "we can gradually find out more about the relationship and possibly the origin and significance of songs" (Nicholson and Koch 1936, 193).

Nicholson and Koch's *Songs of Wild Birds* and its successor *More Songs of Wild Birds* lent themselves readily to the nationwide birdsong inquiry organized by the British Trust for Ornithology from 1937 to 1940 (Alexander 1943).[18] The census included over eighty volunteer observers, who were asked to chart the development of particular species' songs throughout the seasons. Many prewar studies of birdsong had focused on the statistical analysis of singing behavior as part of species' life history descriptions— they quantified and charted the timing and duration of the dawn chorus, average song output per hour, or fluctuations in seasonal song periods. As such, the census had sought to answer existing questions using new data-collection strategies on a much larger scale. By 1940, however, the network was temporarily suspended—initially because of the war conditions but, as it turned out, also due to an unsatisfactory level of accuracy in the results. The data collected suggested that there was enormous variation either in birds' song periods or in the methods of taking down data and the reliability of observers. The statistics led amateur ornithologist Horace Alexander (1943, 67) to conclude rather pessimistically that in reality "not every observer is trained to have an ear that records bird-song heard each day without a good deal of deliberate concentration. The birds may be singing, but the recorder, though within earshot, may not be listening—though he may think he is. Also, some people are much deafer than they think they are."

This observation seems to have been particularly apt for field observers, given the demographics of their community. In 1934, for instance, one ornithologist reported to his peers at the American Ornithologists' Union on the birdsongs he was "losing." William Saunders, who was sixty, had recently had his hearing tested, an increasingly common procedure in the 1930s after Western Electric brought a standardized instrument to market for measuring hearing sensitivity (and the loss thereof) and otologists enthusiastically adopted it. The test had shown his competency on the lower vibrations to be about 90 percent, but his ability "dropped towards zero very fast" for the higher frequencies (Saunders 1934, 504). The results had not come as a surprise, Saunders admitted, as he had been monitoring his steady loss of hearing by comparing his ability to hear the higher notes of specific birds with his younger self and his peers. Specific birdsongs provided a measure whose accuracy was "almost uncanny," even though in some cases his unreliable hearing still proved surprisingly effective. However, Saunders projected, most of the union's members were likely to undergo this trial soon.

At Cornell, Brand reached a similar conclusion in a 1937 paper titled "Why Birdsong Cannot Be Described Adequately," stating that "hearing differs, in all probability quite markedly, from person to person." He pointed out that this usually went unnoticed by the average listener, "but in bird sound the range of frequency is quite different from other common sounds" (Brand 1937b, 11). Brand had personally witnessed a series of experiments at Cornell that demonstrated the existence of considerable "fading points" in the hearing spectra of most people. They also showed that listening had a significant psychological dimension: "We hear what we are listening for and what we expect to hear," for "it is impossible to separate the hearing apparatus from the thinking mechanism" (1937b, 12). Another Cornell graduate and a colleague of Brand's, Harold Axtell (1938, 482–483), found "tremendous variation in the inherent capacity of different people to count notes when delivered rapidly [or] to determine changes in pitch." And "there is the simple psychological factor of suggestion or first association." Upon hearing a bird for the first time, one may be impressed with how much it reminds one of this or that bird. And "because of this initial prejudice, we cannot understand how other people can say that [a song] sounds more like something still different." As a result, Axtell concluded, "bringing our hearing under scientific control seems to be at best an especially difficult accomplishment."

Sound recordings therefore also addressed the problematic nature of listening in a second way. For this fundamental problem with the human faculty of hearing, argued Brand, could now be remedied. On sound film, "extremely short notes and those of very high pitch, often inaudible to the ear, can be clearly seen and studied" (Brand 1935, 40). Although such records were of little help in compiling the long-term data aimed for by the census, they were very useful for documenting and analyzing the composition and acoustic variation of specific songs. Brand's own study of 1935, for instance, highlighted differences in the average frequency of vocalizations in different groups of species, and confirmed that birds produced many more individual notes in a song than a human listener might actually be able to perceive. He presented sound recordings—and particularly sound film—as a "mechanically objective" alternative to the human listener. Apart from teaching the listener, sound recording thus allowed circumventing that listener. Through the careful microscopic study of sound film, birdsong's vibrations, relative loudness, and time distance between individual notes could be studied, allowing "many hitherto unknown facts about birdsong" to be discovered (Brand 1936, 17).

Brand's experiments showed that listeners were usually unable to correctly estimate the number of notes in an ordinary birdsong. Even an accomplished musician and trained listener such as Axtell had required several rounds of listening to estimate that number correctly (Brand and Axtell 1938, 126). Certainly, Brand conceded, a trained ear was still a necessary instrument to hear actual songs when in the field, but he regarded such observations as much less accurate than the ones recorded by the camera. This resulted in a dilemma for the field ornithologist, who was forced to acknowledge that, in Axtell's (1938, 482) words, "what a bird sings" does not equal "what can be heard" by a human observer. Axtell illustrated this point with a detailed description of song features and their variations in the Kirtland's warbler. Writing to aid the observer in recognizing the Kirtland's warbler in the field, he complemented a graphic notation—based on his own hearing—with a graph based on the sound-film recordings he had made of the same song. The juxtaposition of the two graphs demonstrated the relation between an automatic registration of a song's acoustic makeup and its perception by a human listener in the field: although there were striking similarities in shape, the first graph had a much greater resolution of detail. To Axtell (1938, 490), keen to affirm the value of fieldwork, this suggested that while "the human ear is a relatively imperfect recording device and misses many details, it does pick out the essentials."

It was in this ambivalence that electrical sound recording began to redefine what it meant to listen to birdsong. Birds' songs were initially encountered in the field, where listeners, regardless of their listening abilities or musical skill, inevitably relied on "what can be heard." Ornithologists and aspiring contributors made use of auditory field guides and soundbooks as one way to circumvent the problems associated with listening in the field. But as Brand and Axtell's conclusions implied, to study such sounds scientifically also required a closer examination of "what a bird sings." This need for a detailed understanding of a song's acoustic composition pushed the analysis from the field to the laboratory.

Natural Sounds

If electrical recordings allowed sound recordings to be analyzed in the laboratory, the field itself retained substantial legitimacy for these ornithologist-recordists as a space where wild birds could be encountered and their authentic behavior studied. To the ornithologist with an interest in bird vocalizations, Brand (1934, 17) pointed out, "recording the caged wild bird appeared the simpler option" were it not for one major drawback: "Most birds will not sing normally in captivity. Some will not sing at all; others, only in subdued and restrained tones." He added: "We know that molt, willingness and ability to mate, and even incubation are seriously disturbed by confinement; there is every reason to believe that song and song development are also affected," and it would therefore be "rather dangerous to make generalizations on birdsong from observations of captive specimens" (Brand 1936, 49). Brand's Cornell colleague Kellogg, likewise, concluded that "it is doubtful that such recordings [of caged birds] would receive wide acceptance, especially from ornithologists, because it is well known that captive birds behave differently from free birds of the same species" (Kellogg 1938, 175–176). Captive birds were associated with the "tame" and "artificial" rather than with "wild" or "natural" behavior—an idea that harks back at least to early modern natural history (Breittruck 2012).

Yet however well-worn this distinction was among field naturalists, many of the self-styled ethological studies that had begun to appear in the preceding two decades had been based specifically on captive animals. Several of the ethologists concerned had been animal keepers, whose long-term and frequent engagement with the animal, as Konrad Lorenz pointed out, had special advantages over the experimentalist or the field naturalist (Burkhardt 1999). In Germany, Ludwig Koch had made the case for recording wild birds for his first records but was met with derision by his collaborator, ornithologist Oskar Heinroth. Like other pioneers of behavioral study at the time, Heinroth had spent much of his professional career at the Berlin Zoo and in his own aviaries, and in Koch's recollection he insisted that "caged birds always have the same song as members of their own species in their natural surroundings." Heinroth maintained that Koch's "attempt to record birds in the open was merely a sort of circus performance" (Koch 1955, 25–26).[19] Koch eventually managed to convince Heinroth by demonstrating that his caged animals did indeed produce misleading mimicries, with such success that in the introduction to their record *Gefiederte Meistersänger*, Heinroth emphasized that "in no case are the birds recorded caged ones. This kind of recording guarantees a flawless, I might even say, location-specific fidelity to nature [*Naturtreue*]" (Heinroth and Koch 1935, 13).[20]

In the hands of the ornithologist-recordist, qualifications such as the German *naturgetreu* or the English *natural* became complexly layered notions. In the first place, they underlined that the recording had been produced in nature and using real-life specimens. This was certainly not a superfluous assertion in an era when whistling imitators of birds' songs and calls were still commonly featured in the entertainment and recording business and consumers had been familiar only with recordings of caged and trained animals.[21] The distinction between this tradition and nature recordists' commitment to authenticity was not self-evident, given that, as shown in the previous chapter, the artistic imitators that filled record catalogs by Victor, Columbia, or Brunswick portrayed themselves as naturalists and their performances as educational or even scientific. The professional whistler Charles Crawford Gorst, for instance, recorded a series of "Songs and Calls of Our Native Birds" for Victor, but also lectured at the American Museum of Natural History and joined the American Ornithologists' Union, which awarded him a medal for his birdsong interpretations in 1936 (Eley 2014, 12). The ornithologist-recordists' emphasis on the "wildness" and "naturalness" of the birds' environments was thus meant to set the recordings apart from other popular aesthetic repertoires by their scientific-didactic aspirations.

In the context of the move outdoors, though, "fidelity to nature" referred as much to the reproduction's acoustic exactitude and fidelity as it did to its ethological accuracy. Koch himself assured his readers that his "recordings cannot be made in a zoo, but also not in a room or a recording studio, if they are to be really true to life" (Heinroth and Koch 1935, 1).[22] Importantly, what made those indoor places so different from the field for Koch was not only his desire to observe and record authentic behavior: he also considered work with caged birds impossible because "the acoustics of a building distort the sound, sometimes beyond recognition" (Nicholson and Koch 1936, xx). Recordings of enclosed individuals were thus prone not only to distortions in the birds' behavior, but also in its acoustic representation on records. By the late 1920s, consumers had been shaped into critical listeners who were accustomed to expect increasing fidelity from phonographic reproductions.[23] Recordists such as Koch and Brand readily fed into those expectations. Birdsong was very delicate, Brand (1936, 29) wrote, and although the songs had been recorded with great care, ordinary commercial phonographs might easily distort their high-pitched sounds. For that reason, the authors had omitted higher-pitched birdsongs from the record, but listeners would do well to invest in a good phonograph, too. Brand added that the acoustics of the room in which the recording was played were equally important to its fidelity, so it was worthwhile experimenting to find the most favorable place for the phonograph in the room. The particular acoustics might dampen or magnify certain frequencies, and "when sound is so distorted, the reproduction is unnatural" (Brand 1936, 30). For that reason, the Cornell ornithologists often liked to demonstrate their recordings outdoors. Clearly, authentic and undistorted sound mattered for these field recordists. But while a sound recording's fidelity to nature and its acoustic quality are evidently interlocked in this discourse, they proved to be difficult trade-offs in practice.

Mobilizing the Studio into the Field

Despite the advantages that the new media technologies were expected to afford recordists in their analysis of animal vocalizations, and the importance of the field to this endeavor, putting such technologies to use effectively in the field proved a daunting task. Mobilizing electrical recorders and making them work in the field site turned out to require more than the usual degree of improvisation and make-do.

When the intensity of their collaboration with the movie industry began to wane, the recordists at Cornell soon found that they lacked the practical knowledge needed to assemble the right equipment and make it work: "The whole thing seemed so complex we didn't know where to begin. Like an octopus it seemed, and each of its tentacles reached out into a different field of technology: mechanics, electrical engineering, sound and acoustics, photography … all seemed hazily intermingled in this new machine to which we were being introduced" (Kellogg 1938, 78–79). Only after a full year of testing and tinkering with the assistance of their colleagues from the electrical engineering department did the ornithologists eventually manage to redesign their "rather amateurish" outfit so that it would be able to withstand hard usage in the field and "be capable of being operated by ornithologists who knew little or nothing about engineering" (Kellogg 1938, 84).

This work reduced the technical complexity of the recording enterprise, but actually recording in the field remained a logistical tour de force, even for seasoned recording engineers. "A layman might think it easy, and might imagine setting out with a small apparatus and a microphone like an amateur photographer with his camera, having no more to do than to take a snapshot," Koch noted in a chapter on the production of sound records (Nicholson and Koch 1937, xx), but in reality, field recording was much more complicated.

To record birdsongs in the field, recordists relied on their familiar studio equipment, which they transported to the location in a large van (figure 3.1). The van was outfitted in such a way that recordists referred to its improvised control paneling as their own mobile studio. In what was perhaps a tongue-in-cheek remark, Nicholson and Koch (1937, 4) judged it to be extremely similar, except in size, to the professional studios at Abbey Road. The vans carried the amplifiers and multiple microphones, as well as a great number of dry-cell batteries or a noisy dynamotor to generate power on the spot (Pulling 1949). All this did not, however, preclude frequent failures. The power supply was often inadequate, and humidity regularly caused short circuits. Brand and his Cornell colleagues moved a large and extremely sensitive sound camera around, which together with the testing equipment amounted to a weight of roughly 1,500 pounds, causing dangerous sways of the van. Instead of the sound camera, Koch and his British recording engineers worked with a phonograph recorder that could cut the recordings on wax discs. Recording on wax had the disadvantage that

Figure 3.1
Preparing and leveling the recording van in the field.
Source: E. M. Nicholson and L. Koch, *More Songs of Wild Birds* (London: H. F. & G. Witherby, 1937), viii.

the recording could not be played back on location but had to be processed in the studio first, requiring great care to prevent breakage of the discs in the meantime. Moreover, wax could not be recorded on in its original state: the discs first had to be softened to a critical point in special wax heaters or large electric ovens. To make the needle run smoothly on the wax, the recording vans also had to be absolutely level so that the equipment would stand perfectly upright.

To get the studio on-site and operate all the equipment smoothly, Nicholson and Koch relied on up to five staff members, including a driver, technician, and recording engineer, who brought, set up, and operated the equipment on site. At Cornell, the ornithologists typically managed these tasks with just two operators. The recordists often started their field operations well before sunrise to catch the dawn chorus, when birds are most active and vocalize most forcefully and frequently. Koch's team, for instance, would set off into the field at 3:00 or 4:00 a.m., checking the microphones they had positioned there the day before to allow the bird to return to its perch overnight. With their equipment connected, checked, and prepared, the operators had to wait for the bird to begin its song. An observer would typically coordinate with the recording engineer over a separate line, to allow him to lower the stylus onto the recording medium at the exact moment (figure 3.2). Often, however, the disturbed bird would have flown off, often out of range of the stationary microphones. With his artisanal notation techniques of pencil and notebook, the field naturalist had no trouble following his subjects as they moved between their perches—but the ornithologist-recordists, dependent on heavy equipment, were less flexible and hence both less mobile and more intrusive. Recording the birds as they sang, therefore, required a great deal of patience, coordination, and luck.

Historians of field research have recently noted the importance of infrastructure such as railroad lines for the practice of scientific fieldwork,

Figure 3.2
Listening at the loudspeakers in the mobile studio. Koch sits to the right.
Source: E. M. Nicholson and L. Koch, *More Songs of Wild Birds* (London: H. F. & G. Witherby, 1937), 5.

expanding access to distant field sites and facilitating the transport of equipment, specimens, and personnel between center and periphery.[24] This was certainly the case for birdsong recordists. Whether their destination was the cultured environment of town parks or the relative wilderness of nature reserves, they relied on roads, paths, trails, and other infrastructure as a network of mobility that became vital to the logistics of nature recording. Some recordists were able to employ this infrastructure creatively— initially, radio recordists such as Carl Weismann in Denmark and Sture Palmér in Sweden each started recording wildlife sounds next to railways, which allowed them to use the telegraph wires to send the recording signal to a wax-disc recorder in their respective radio headquarters.[25] Just as often, however, the naturalist-recordists found such networks limiting and insufficient in range. Their subjects rarely kept to the immediate surroundings of roads and paths, forcing the recordists to venture into much wilder terrain. Although they often managed to buy themselves a wider range of operation by making the cables between microphone and recorder up to a mile long, this did not always suffice. During the expedition by the Cornell ornithologists and the American Museum of Natural History described at the beginning of chapter 1, the ornithologists were forced to transfer their sensitive recording equipment from the studio truck onto mule-drawn carts. It took them two days to dismantle the equipment from the truck and mount it on the carts. This turned out to be the only possible mode of transportation to navigate the extensive Southern swamplands.

Even when field recordists were able to stray from their facilities and recording conditions became far from optimal, the field was never as much a wilderness as it might have appeared. As Brand (1932, 438) warned his readers, "It is not as simple as it would seem to get a location where there is absolute quiet. ... Too great proximity to a traffic road, for instance, makes recording impossible." Ironically, it was precisely the infrastructure required to reach distant areas for recording that produced the noisy surroundings. The Cornell recordists discovered this the hard way, for instance, when several of their hard-won recordings turned out to be useless because the microphone had picked up static from nearby power lines. Both Brand (1932) and Koch (1955) complained that their recordings of bird sound were often interrupted by a passing airplane or compromised by the hum of a distant highway. Conservationist Max Nicholson accompanied Koch on several expeditions into the British countryside, and vividly recalled his impressions there:

Aeroplanes, motor-cars, lorries, trains, and motor-bicycles combined to shatter the tranquility which had been so perfect a few hours before. Just as smoke pollution helps to swamp a town under fog, so the natural peace of the country was drowned under the indefinable hum of distant engines and wheels. ... Until one has listened objectively to all these sounds coming through the loud-speaker, in what counts still as a peaceful retreat from the bustle of London, it is hard to realize what a noise-ridden world we have managed to make ourselves live in. (Nicholson and Koch 1937, 38)

As a trope that was commonly employed in noise-abatement campaigns from the late nineteenth century onward, the analogy between noise and smoke pollution is suggestive of these noises' ungraspable omnipresence. Indeed, the recordists' publicly aired aversion to noise matched a more widespread resistance to all kinds of mechanically generated noise, which, as historians of technology have shown, became increasingly organized in American and European cities between 1910 and 1940.[26] In New York, for instance, officials of the Noise Abatement Commission began monitoring sound levels in the streets in 1929; in London, an Anti-Noise League was formed in 1933. The league framed noise as a threat to citizen health and, with the establishment of the decibel as a unit of loudness in 1925 and the availability of technical instruments such as acoustic meters, claimed a more "systematic and objective measurement" of city noise levels. But the battle against sonic pollution was not restricted to the clamor of the city. In the English countryside, noise became a point of concern too. Thus, in 1936, the geographer Vaughan Cornish argued on behalf of the Council for the Protection of Rural England (CPRE) for better regulation of the Norfolk Broads region, deploring the vulgarity of anthropogenic noise. Complaining about the gramophones and loud behavior of the holidaying masses, he pointed out that "it is well to remind ourselves of the fact that quietude and harmony of sound are among the amenities of the natural scene" (quoted in Matless 2005, 752). Nicholson's complaint about advancing noise levels in the countryside, it seems, matched a sentiment that was felt more generally.

Nicholson's objection was not just sentimental, however. In this mechanical age, the soundscape had obviously changed. This was not simply because mechanical artifacts and machinery generated more noise, but also because measuring devices and recording technologies made noise increasingly audible. Such technologies had begun to structure the sonic landscape in a way completely different from the way the human ear had done. The recordists attributed this above all to the unselective sensitivity of the microphone: "The sensitive microphone takes up all noises, often within several miles' radius, and exaggerates them" (Nicholson and Koch 1937, 20). As a result, an observer such as Nicholson would often find that "although I did not realize it at the time, an aeroplane was cruising about too far away to be seriously noticeable to the human ear, but quite near enough for the microphone with its acute sensitivity to pick up the sound" (Nicholson and Koch 1937, 36). Microphones, Nicholson's American counterpart Brand (1934, 19) explained, picked up "ground noises":

> These are the sounds that are always present, but which our ears with the aid of our brain shut out from our consciousness. The ticking of a clock in our bedroom does not disturb our sleep—we say we become accustomed to it. Our brain does not transmit to us all the sounds which our ears pick up. But though man has been able to invent a mechanical ear—the microphone, and it is an extremely sensitive one, too—he has not been able to equip it with a brain. It transmits all the sounds it hears, without discrimination.

The "mechanical ear," then, did not act as a simple prosthesis to the human ear; it did not simply amplify what a human ear would normally perceive. It also structured the listeners' acoustic experience in a new way, through its indiscriminate, unselective registration. While the mechanical ear promised greater objectivity, it also came at a price. Without selection and idealization, images become cluttered with the noise of context and artifacts (Daston and Galison 2007). For the naturalist-recordists, context imposed itself on their objective registrations in the same way, with ambient sounds heard as equally unpleasant noises.

Such noises posed an aesthetic problem for recordists intent on publishing their records commercially. But they were also an epistemic concern. If the hum of distant engines threatened to compromise the evocation of wilderness that record producers like Koch were so keen to construct, it was even more disruptive to the scientific analysis of the sound-film recordings that Brand and his colleagues sought to advance. They had established that the human ear would usually be able to select and distinguish a bird sound relatively efficiently from its noisy surroundings, but once the song was represented visually on sound film, the distinction between sound and noise became potentially harder to make. The variable-density method of sound-film recording that Cornell recordists used allowed frequencies to be distinguished by the distance between very thin hairlines that were "hardly visible to the eye." The closer together such lines were on the film, the higher the pitch. When these lines overlapped with the frequencies of other noises, they became especially difficult to distinguish. I will return to these concerns regarding the visualization of sound in chapter 5. At this point, however, it is important to note that although sound-on-film inscriptions offered ornithologists new ways of comparing and a claim to objectivity, the process of visualization also conditioned what kinds of sounds could be recorded and studied in the first place. With its sensitive microphones and recording devices capable of capturing a wider frequency range, electrical recording had seemed to herald a solution to the problem of faithful reproduction of sound, but it simultaneously created a new problem: the acute perception of noise, which the Cornell recordists believed might complicate the visual inspection of their sounds for scientific study.

Controlling Field Acoustics

The recordists in Britain and at Cornell University faced a similar challenge. Their commitment to recording authentic, natural behavior had led them to record in the field, yet the very authenticity of the field site, with its abundance of ambient noises, complicated the process of recording and subsequent analyses. Such disturbances confronted recordists with the question of how this soundscape could be authentically reconstituted on a recording. Their responses differed quite markedly.

Koch and his recording team tackled the realities of field recording by employing a strategy common in the studio recording practice of the day. They engineered the desired sound by adjusting sound levels through a strategic placement of microphones and the singing bird. Historian of sound technology Susan Schmidt Horning (2013) has argued that much recording work in the first half of the twentieth century was based on trial and error and the gradual acquisition of experience. Before the advent of electrical systems able to amplify the recording signal, studio recordists relied on their intuition to position the vocalist and the various instruments around the recording horn in perfect balance. In fact, this remained the case even when the condenser microphone and vacuum tube amplifier replaced the recording horn in the 1920s, with a dramatic improvement in sound quality. By the late 1930s, the newly named "recording engineer" had much more control over the process of recording. This did not, however, reduce the need for great skill and experience, particularly in placing the increasingly sensitive microphones in a spatial relation to the sound source—expertise that could only be acquired through trial and error. It was this particular skill that the recording engineers and technicians in Koch's unit had learned in the studio. Except for Koch, who had recorded outdoors and in zoos, these men had little or no previous experience in recording wildlife outdoors. In tackling the acoustic complexity of the field site, therefore, Koch and his team outlined an approach similar to that used in studio recording. Koch (1955, 3) explained that a real birdsong hunter must "with skill and much patience go about the job of bringing the songster as near as possible to his microphone, or rather the microphone as near as possible to the songster." At all times, Koch's (1955, 73) recordings were created by arranging the microphone to capture the singing bird "as if he were a performer in a studio."

Similar as it was to studio recording, the technique differed in at least one important respect. By the 1930s, the space of the sound studio had begun to be operated as a "dead room," an acoustically sterile backdrop for musical performance (Schmidt Horning 2013; Thompson 1997). In contrast, the British recordists elevated the natural environment, with all its "noise," to a key component of their kind of nature recording. As Julian Huxley pointed out in his laudatory introduction to Koch's first British soundbook, Koch's records had "the quality for evoking the bird's environment": "When I first heard his records I was immediately struck by the way in which they called up the natural environment of the singers. As the nightingale's voice escaped from its ebonite prison under the touch of the needle and the scientific magic of the sound-box, I felt myself transported to dusk in an April copse wood" (Nicholson and Koch 1936, xiii–xiv). This "natural" sound, the evocation of an environment, was exactly what Koch was after: "In my nature recording I invariably observe the principle of getting the bird-notes in question in the foreground by all means, but always with their natural background" (Nicholson and Koch 1937, 21).

Under such circumstances, evoking a sense of naturalness required a number of crucial interventions. Most importantly, the team had to be sure to distinguish between the bird's voice in the foreground and the interfering and sometimes noisy environment in the background. To do so, the recordists would arrange up to six different microphones to enclose the bird's song perch, improvising as it were a large-scale recording room outdoors (figure 3.3). Each of these microphones was linked to a control panel in the van several hundred yards away, where an incoming sound to one of them would be cut straight into the wax disc—a setup that was not used to combine multiple signals into a stereophonic recording, but merely enhanced the chance of registering a bird's "performance" close enough to the microphone to pick up its signal without too much ambient noise. Occasionally the bird would fly away, and then the whole setup of microphones would have to be changed, including tests to make sure the microphones "cut out as much as possible of interfering noises" (Nicholson

Figure 3.3
Cornell ornithologist James Tanner aiming a parabolic reflector.
Source: Albert R. Brand Papers, #21-18-899, Folder 2:1, Division of Rare and Manuscript Collections, Carl A. Kroch Library, Cornell University.

and Koch 1937, 37–38). The sensitive omnidirectional microphones meant that disturbance could be caused by any ambient sound: the low frequencies of anthropogenic and mechanical noises, electrical interferences, and the constant or sudden interference of rain, wind, water, rustling leaves, or other species. Although the recordists considered noise part of the reality of field recording, the intended purpose of the recordings—reproduction on low-tech gramophones or radio broadcasts—also dictated that interference be reduced as far as possible by obtaining a close-up recording of the bird. Of course, the very act of recording always implies mediation and selection. In this context, therefore, the natural soundscape of the field was a sound produced by technical restrictions of future reproduction and an implicit preference for how nature should ideally sound—tranquil and clear.

The Cornell recordists faced similar problems, but eventually they took a rather different approach. After experimenting for some months, the ornithologists experienced some success with their equipment, but also found that many of the recordings they had produced with ordinary microphone setups had been significantly impaired by ambient noise. With some tinkering and re-recording, they had learned to filter out low-frequency sounds to an extent. However, this tended to result in a high-pitched and squeaky tone, making it immediately obvious that the sound had been "worked." Given their stress on the faithful reproduction of natural sounds, this was an effect the recordists were determined to avoid. The group found inspiration for a solution in a *Radio News* cover story on a parabolic reflector and a suggestion by their colleagues in the physics department. "Sound concentrators" in the curved shape of a parabola had been patented for radio pickup only a few years before and were being used extensively in sports broadcasting, movie productions, and even studio recording (Dreher 1931; Elway 1932; Schmidt Horning 2013; Thompson 2002). The group cast their own parabola from an old mold that had been used for wartime airplane detection. Reviewing this innovation with the benefit of hindsight, Kellogg (1962a, 39) concluded enthusiastically that the "parabola for picking up and concentrating birdsongs on the microphone was probably our greatest piece of good fortune. Recording would have been possible without a reflector, but the results with it were so superior as to make the instrument a universal tool in this field."

For the Cornell ornithologists, the source of the parabolic reflector's superiority was that its surface reflected sound waves to a dynamic microphone at its focal point, enormously increasing the input to the recording equipment and concentrating it to at least 20 dB louder than the sounds not caught by the parabola. This amounted to an amplification of about fifteen times (Sellar 1976). As a result, the reflector could help to record sound at much greater distances than microphones in any conventional setup. This provided field recordists with plenty of advantages: bird sounds could now be recorded at a greater distance, in flight, or in places inaccessible to recordists using traditional techniques of microphone placement— an obvious added advantage to the ornithologists, who were becoming increasingly interested in studying animal behavior in the wild (figure 3.4). Since microphones would not have to be stealthily set up, rearranged, and retrieved, recording was much more efficient too.

Equally important was the fact that the parabolic reflector only focused the sound energy from the direction in which it was pointed, which permitted greater selectivity and directionality in recording. Unlike the common carbon or condenser microphones used by recordists more generally, which recorded omnidirectionally, this microphone design meant that "the outside noises are very nearly shut out, while the sounds wanted are greatly increased" (Brand 1934, 23). These advantages led Kellogg, Brand, and their colleagues to compare the concentrator "with the magnifying power of a field glass, reducing the apparent distance to about 1/20th the actual distance" (Kellogg 1938, 181). The analogy to this quintessential field naturalists' tool was particularly apt, and not just because the reflector had a small telescope attached to bring the individual bird "into focus." It also fit because just like a pair of binoculars, the concentrator would eliminate some parts of the (aural) landscape but zoom in to produce a close-up of others. As with a visual lens, then, this aural lens produced a close-up recording. This was a sonorous sound with a high signal-to-noise resolution, which provided very little information on the spatial behavior of its source signal: it cut out part of the reverberation, for instance. In this sense, close-up sound resembled the skillfully engineered sounds being produced in the recording studios of the 1930s, capturing the sound and nothing but the sound. Of course, bird recordists did not have access to the same modern means of architectural control of sound as professional studio engineers. But they did share with the engineers a desire to produce an attractively intelligible sound that, to paraphrase film theorist Rick Altman (1992, 61), privileged the listener as a consumer of sounds.

Despite the similarity in kind between the studio and the new field recording techniques, declarations that the sound concentrator had "overcome much of the handicap imposed by lack of soundproof studios where the wild birds sing" come across as somewhat hyperbolic ("Reflectors like Airplane Detectors Catch Bird Songs," 1934). While it was certainly true that

Figure 3.4
British recordists setting up a microphone near a song perch.
Source: E. M. Nicholson and L. Koch, *More Songs of Wild Birds* (London: H. F. & G. Witherby, 1937), 4.

the parabolic microphone was much more effective in capturing birdsongs in the field than any other type of microphone, the recordings considered most satisfactory by the Cornell ornithologists were mostly achieved on quiet days and at a relatively short distance. In acoustically less favorable surroundings, interferences still occasionally leaked through. Furthermore, although the parabola excluded most ambient noises and distortions, it was itself a source of new interferences. The parabolic recordist focused the microphone manually rather than by monitoring a fixed microphone at a distance. As a result, handling noises and unintentional contacts with the large plaster shape were easily picked up. An additional disadvantage was that, due to its design, the concentrator structurally suppressed lower frequencies, whose larger waveforms were only partially reflected by the parabola, changing the quality of the recorded sound.

Not all listeners evaluated these effects the same way. Recording technicians unfamiliar with birdsong usually preferred to cut out a wide band of lower frequencies to reduce noise levels, whereas the "sensitive ears" of listeners more experienced with the intricacies of bird vocalizations, as Kellogg complained, found that this seriously detracted from the experience of the songs. The "absence or suppression of low frequency components" made the songs sound thin and squeaky (Kellogg 1938, 183–184). Ornithologist-recordists were thus faced with a difficult choice between a demanding recording routine that produced noisier recordings of a more complete acoustic quality, and a more efficient approach that produced focused yet subtly distorted acoustics. Although recordists agreed that not all noise detracted, by definition, from the scientific and aesthetic potential of the recordings, extensive listening tests among the Cornell recordists ultimately led them to favor the parabolic microphone as their standard tool.

Each in its own way, the recording techniques used by Koch and Brand fundamentally aimed to achieve a degree of control over the acoustic environment in which birds' songs were recorded, by enabling recordists to differentiate between the sound levels of wanted and unwanted sounds. As such, they helped the recordist reconstruct an idealized and authenticated version of the soundscape from the very start, when recording in the field. However, strategic microphone placement and parabolic amplification, as technologies to order and represent the soundscapes of the field, also differed in crucial ways.

One of those differences is captured in two schemes in a 1952 recording manual for aspiring recordists (figure 3.5). It places the ordinary dynamic microphone in the middle of a circular recording range, suggesting the

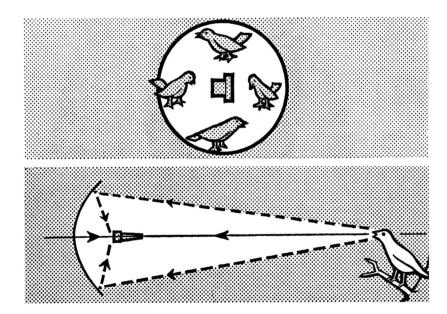

Figure 3.5
Schemes illustrating the recording range of an ordinary crystal microphone (top) and
the parabolic reflector (bottom).
Source: F. Purves, *Bird Song Recording* (London: Focal Press, 1962), 38.

microphone's unselective reproduction of the aural events within its perceptual scope. The parabolic microphone, on the other hand, is portrayed
as exclusively capturing the sounds emitted by a single bird, and nothing
else. This scheme makes clear that the two recording techniques constitute
different sonic topographies in the representation of the field site. Because
directional microphones amplified especially those sounds at which the
reflector was aimed, they did not permit a great deal of local acoustic context
to be recorded. Nor did they evoke the bird's environment with what British biologist and science popularizer Julian Huxley appreciatively described
as the "fullness, immediacy, and emotional completeness" of aural immersion he had experienced when listening to Koch and Nicholson's British
bird recordings (Nicholson and Koch 1936, xiv). This is not to suggest that
other types of recordings could not sound sonorous or close-up, or that the
parabolic effect was always clearly discernible. What it does show is that by
its very design, the parabolic microphone favored close-up recordings and
permitted sequences of acoustic behavior to be sampled out of a panoramic,
enveloping, and variously noisy aural ecology.

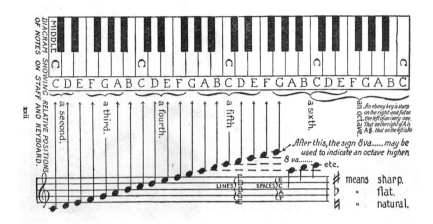

Plate 1

A musical key for readers "who do not read music" by Ferdinand S. Mathews.

Source: F. S. Mathews, *Field Book of Wild Birds and Their Music: A Description of the Character and Music of Birds* (New York: Putnam, 1904), xxii.

Plate 2

Cover of *Songs of Wild Birds* by E. Max Nicholson and Ludwig Koch.

Source: E. M. Nicholson and L. Koch, *Songs of Wild Birds* (London: H. F. Witherby, 1936).

Plate 3

Front cover to *A Field Guide to Western Bird Songs*.

Source: P. P. Kellogg, *A Field Guide to Western Bird Songs* (Boston: Houghton Mifflin, 1962c). Three 33⅓ rpm discs. Reproduced with kind permission of Houghton Mifflin.

Plate 4
BBC Sound Archives Natural History Recordings Catalogue.
*Source: BBC Sound Archives Natural History Recordings Catalogue. Supplement to Volume
One: Species.* British Broadcasting Corporation, 1978. Courtesy of Jeffery Boswall and
the British Broadcasting Corporation.

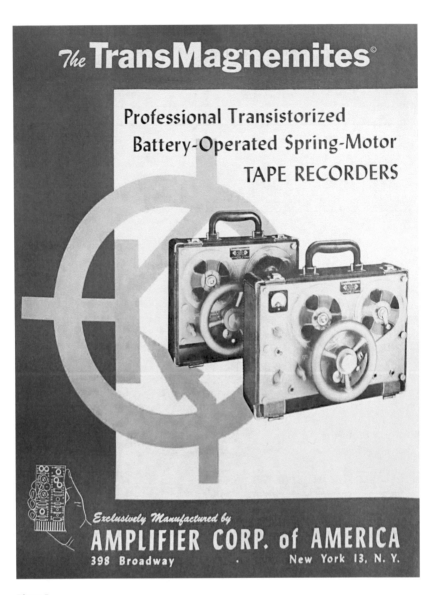

Plate 5

Advertisement for TransMagnemite magnetic tape recorders by Amplifier Corporation of America. An improved version of the Magnemite portable magnetic tape recorder that Peter Paul Kellogg helped design and that went into production in 1951.

Source: Peter Paul Kellogg Papers, #21-18-893, Folder 4:6, Division of Rare and Manuscript Collections, Carl A. Kroch Library, Cornell University.

Ordering Landscapes

The effect of the microphone design on the soundscape of the field was not just aesthetic—it also had epistemic implications. The microphone throws into relief some of the selection processes that science studies have shown to be at work when observations are crafted into scientific objects. Such processes embed the parabolic recorder into the field's laboratorization. To understand how a technology such as the parabolic microphone could reorder an aural landscape and what epistemic advantages this granted the listener, it is helpful to trace an analogy with the technological mediation of the landscape in the visual domain. Such visual strategies for representing, reproducing, and ordering a complex natural environment, whether artistically or scientifically, provide useful reference points for understanding how directional recording began to order the field's soundscape in new ways.

Environmental historians have compellingly shown how pictorial conventions mediate different relations between nature and its inhabitants, and in this way produce different landscapes. This becomes particularly evident in the instrumentalization of the camera by various field sciences. For some field biologists, for instance, the camera was a means to introduce audiences to the individuality and private lives of animals in nature and to convey the emotional essence of a unique experience with nature. Others applied the camera to produce a wide-angled, panoramic field of observation that instead distanced researchers and viewers from their objects of study (Mitman 1996). This distancing effect of the camera is amplified even further in aerial photography, which reproduces the same landscape as a unified panoramic view from above or as a fragmented collection of analyzable "units" (Kwa 2008). This distancing, godlike perspective "from nowhere" is neither a property of the landscape nor merely a presupposition of the viewer. Rather, the camera stands between nature and its representation, with its own history of innovations and associated ways of looking. Angle, distance, and scale reproduce the landscape within a two-dimensional space. But as landscape painting reminds us, in so reproducing the landscape, the observer also takes charge of it. In an accepted account of nineteenth-century landscape painting, painters are said to have freed themselves from studio conventions and to have begun painting "real landscapes" instead. But even when landscapes were painted outside the studio, their work was never free from the conventions guiding studio representation (Alpers 1998). Like the laboratory, the studio is not just a physical workplace, but also an instrument and a condition of work that make it possible to manipulate the realities of natural objects. In that frame, land-scapes, too, can be represented with the static and frontal presence of a nearby object, as a still-life object.

Such visual technologies of landscape representation help us under-stand the complex ways in which directional, parabolic recording orders the acoustic realities of the field. The microphone's effect is twofold. First, by selectively amplifying the sounds at which it is directed, the parabolic reflector introduces a new aural perspective. Physically, the microphone allows recordists to take up their position at a distance from the bird, and thus reduces the need to intervene and disturb its behavior. Aurally, in con-trast, it generates a frontal and almost perspectiveless close-up recording that seems to bring listeners into extremely close proximity to the object of recording—the bird. The technique thus preserves the spatial and phys-ical relationships between observer and observed in the field, but collapses those relationships in the listeners' experience. For listeners, an experience of aural immersion makes room for a detached, objectifying point from nowhere that brings them very close but situates their presence nonethe-less outside the scene of action. Second, even if directional recording sug-gests an intimate closeness to the individual bird, it also allows the field to be reproduced as a set of fragmented cuts. By focusing and intensifying particular segments of a soundscape, it favors, more than any other con-ventional recording technique, an aesthetic of the individual as a standard unit of analysis. The parabolic microphone discriminates songs of individ-ual birds not just against a background produced by anthropogenic activ-ity, but also against the noisy ambience of their natural habitat or social activity in interaction with other species (bird or otherwise). It is this seg-menting effect that enabled individual sounds to be collected, labeled, and ordered—in short, to be sampled from a complex acoustic ecology—much like any other type of natural history specimens.

Together, these effects allowed the recordist to reproduce a living bird as an aural analogy to the conventional images presented in contempo-rary birdwatchers' field guides such as Roger Tory Peterson's *Field Guide to the Birds*, whose first edition was published in 1934: a decontextualized specimen with a minimum of reference to naturalistic surroundings, situ-ated against an opaque, hazy, or otherwise cropped backdrop and repre-sented without the conventions of linear perspective (Law and Lynch 1988, 279–286). The images assume their own sense of realism—one that highly idealizes what can be heard in the field itself and that present-day recordists describe as a shift in "perspective," from the birdwatcher's situated posi-tion, experiencing the bird in the field, to that of the singing bird itself.[27]

Sounds, in other words, are reproduced as if they are experienced by a transcendent ear-from-nowhere. Without such a fixed point of hearing or the suggestion of perspective, the aural gaze fixates the bird as a still-life motif in a clearly demarcated acoustic landscape. In this sense, the parabolic microphone also connects with a much broader set of modern orientations toward sound and listening that crystallized around new technologies of sound mediation such as the stethoscope, the telegraph, and the headset. According to media historian Jonathan Sterne (2003, 93), for listening to become useful as a tool of rationality, hearing had to be separated, intensified, and focused as a discrete activity, an "audile technique," as he calls it. Such audile techniques actively transformed acoustic space for the purposes of listening; by segmenting, cutting up, and reassembling sound, these techniques articulated an auditory space that became amenable to close scrutiny and specialized description.

It is in this context that we may revisit the analogy between the art studio and the laboratory as instruments of investigation, as a starting point for considering the ways media bring expansive environments under control. Not just a space, the studio becomes a frame for analyzing and investigating (rather than imitating) a landscape. In the same way, the laboratory does not remain exclusive to an experimental culture within the laboratory walls, but may instead impinge on the techniques of fieldwork itself. It is in this same way, I propose, that the parabolic recording technique developed by the Cornell recordists may be understood as enforcing a laboratory condition in the field. The parabolic reflector here is a technology connecting the schematic opposites of field and laboratory.

How exactly did this recording technique contribute to the laboratorization of the field? The field site remained a place where animals could be studied in their natural environment. It seemed to permit access to their vocal behavior in an unadulterated state, *as it was found*, as opposed to the perceived artificiality of the prepared, tamed, caged, or otherwise domesticated bird. This is the reason that birds could not be recorded in the studio. Natural environments appeared to produce natural behavior and acoustic conditions; on the other hand, they were difficult to control. To make birdsongs amenable to recording, measurement, and analysis, they had to be domesticated—acoustically more than physically—by isolating them from their environment. This is not to say that locality and context did not matter at all. In fact, locality and the context of a place mattered a great deal to ornithologists, who, as chapter 5 details, were keen to demonstrate behavioral variations. Context was ordered through extensive field notes that demanded and privileged descriptions of the unique local circumstances of the field and recording site. But at the same time, local acoustic conditions were effectively fashioned into a generic, homogenized, and antiseptic background that favored a detached aural presence over immersive experience. This sterilizing process enabled the laboratorization of the field.

The field thus underwent "laboratorization" in several ways. First, directional recording techniques reproduced the bird's vocalizations against a generic background sound that was, at least in ideal terms, similar to the expectation of a generic soundscape within the isolated laboratory. Even as an acoustic environment more complex than that of the laboratory, then, the field is presented as a place of science in which a sterile and mute backdrop provides the conditions for knowledge production to take place. Second, the laboratory stands for a particular set of epistemic techniques by means of which, as a number of classic laboratory studies in the sociology of science have shown, scientists produce "docile objects."[28] The sequential transformations that Michael Lynch (1988) has shown to be involved in rendering an original image into a scientific representation also apply to the process of sound recording in the field. Just like illustrations, diagrams, and pictures, aural inscriptions too are selected, filtered, isolated, and reduced. As such, they reproduce the field site as a specific phenomenal field. This phenomenal field allows a landscape to be transformed into a graphic space by imposing the plane geometry of paper onto the field site, making phenomena in the field site amenable to later mathematization, calculation, and a cascade of further inscriptions. Of course, birdsong recordings, as audible "inscriptions," are not yet subjected to mathematization at this stage, in the field, but they will be eventually, as chapter 5 shows in more detail.

Hence, third, the soundscape of the field has been "laboratorized" in the sense that these techniques ultimately facilitated the integration of field recordings into laboratory processes and the production of new kinds of inscriptions. In this very concrete sense, too, reproductions of the field were shaped by the practical demands of laboratory analysis: the microscopic analysis of sound film advanced by Brand and his Cornell colleagues benefited from focused, compartmentalized, and sound-sterile recordings. Recordings with simultaneous voices by multiple individuals or species, for instance, did not translate well visually, especially when different voices covered the same frequency band. Imposing a new topography on the field was therefore also instrumental in ensuring the subsequent intelligibility of recordings as visual inscriptions in the laboratory.

Like the laboratory model in the nineteenth century, the parabolic reflector traveled widely, beyond the local and historical context of the Cornell group of ornithologists. Koch and his fellow recordists long resisted the parabolic microphone on aesthetic terms, but by the early 1950s a change of curator in the BBC natural sound archive overturned this stronghold of traditional outdoor "studio" techniques and popularized the device among British recordists and birdsong biologists (Marler 1956; Simms and Wade 1953). Indeed, Cornell ornithologist Peter Kellogg was hardly exaggerating when, in 1962, he described the reflector as a "universal tool" in the field (Kellogg 1962a). By then, the parabolic microphone had been adopted as a part of the standard outfit of almost every amateur and professional recordist engaged in nature recording. In subsequent decades, British archivists promoted the parabolic reflector among recordists as the most efficient way to undertake nature recording (Margoschis 1977; Purves 1962; Sellar 1976). More recently, its clumsy design has been complemented by other, sometimes more effective types of directional recording, such as the "shotgun" microphone models (Catchpole and Slater 1995, 13–14). The parabolic reflector's success in becoming a "universal tool" in the field was due in part to its relative simplicity, which made it affordable for a large group of nature recordists, and to its maximization of the time spent recording in the field. As chapter 4 will show, its success was also due to the Cornell ornithologists' own strategic policies. But above all, the parabolic reflector traveled so widely because unlike the soundproof rooms that ethologists built (to experiment with birds' song learning), it had the benefit of portability and little or no institutional investment (Thorpe and Hinde 1956). It also provided a powerful way of ordering a complex acoustic ecology into units of individual songs—which could subsequently be repackaged as sound effects, published records, or scientific analyses—in the same way that visual technologies ordered landscapes.

Analogies with visual tools and metaphors such as "close-up" or "focus" presented themselves easily, in fact, especially for developers and users of the parabolic reflectors themselves. This is not coincidental. What may, at first glance, seem to illustrate the dominance of a visual discourse in Western culture, language, and science (Classen 1993; Levin 1993) may instead be read as evidence of how visual and audile techniques organize perception in roughly analogous ways and have similar epistemic effects. Like audile techniques such as the stethoscope, the parabolic reflector highlights an affinity between cognitive and epistemic effects across the presumed sensory divide in which hearing is considered an immersive, subjective, and affective faculty that places the individual in close contact with the world, while vision, in contrast, is generally outwardly directed, directional, distancing, abstracting, isolating, indifferent, and by those virtues, ultimately also objectifying (Sterne 2003, 15). The history of the parabolic reflector should thus be read as a warning against any conception of sound and vision that asserts essential, transcultural, or transhistorical features over the historical, local, and practical organization of perception.

4   Sampling Assets: Economies of Scientific Exchange at the Cornell Library of Natural Sounds

Sharing Sounds

In April 1956, Pennsylvania State University hosted some forty-five inter-national scholars who had convened on the initiative of Penn State pro-fessor Hubert Frings to form an International Committee of Bioacoustics (ICBA). They represented the few laboratories in sensory physiology and animal behavior studies worldwide that had recently turned to studying animals' acoustic communication, and Frings had invited them to Penn-sylvania to think about the future. The past two decades had brought sig-nificant advances in electronics and acoustics, he noted, but while this had resulted in powerful tools for the production, analysis, and recording of sounds, little exchange of material had taken place among the new users.[1] Frings suspected that a wealth of acoustic data therefore lay dormant in bioacousticians' offices and laboratories, unchecked or simply untapped by peers.

To solve this problem of access and exchange, the committee proposed to establish two new institutions: an International Library of Publications in Biological Acoustics and an Animal Phonography Collection, which was to collect biologically significant samples recorded by professional bioac-ousticians. Yet with physiologists, ecologists, taxonomists, and ethologists around the table, no one seemed quite certain either which sounds were significant or exactly who should be the judge.[2] Nor was it clear from the outset how and where the recordings would be collected, though one of the participants had his own ideas on that point. Cornell professor of orni-thology and bioacoustics Peter Kellogg told the committee that the foun-dations for a truly international center for biological acoustics had already been laid at the Cornell Laboratory of Ornithology, where he maintained a catalog of natural sounds. After the death of Albert R. Brand in 1940, an endowment had initially provided for the establishment of a modest

"Bird-Song Laboratory" at Cornell, where Kellogg pursued his recording work with his colleague Allen. Their collection of animal records having expanded, Kellogg explained, he wished to rename it the "Library of Natural Sounds," an institute that could well serve as the discipline's central data repository.

The meeting eagerly accepted Kellogg's offer to archive any acoustic sample that members submitted, along with their published papers for reference—Kellogg anyway expected the sounds to fit on a single reel of tape. But to his frustration, he failed to win the participants over to the more ambitious plan: his vision of building a complete and international scientific sound collection at Cornell. At the close of the committee meeting, the proposed scientific archive of bioacoustics samples existed only on paper. Although the published proceedings did much to raise the profile of the new discipline of bioacoustics and generated national offshoots such as the Committee for Biological Acoustics in Britain, the ICBA never acquired sufficient financial backing to survive.[3] Unsure of the collection's status, members also hesitated to submit their material to Kellogg, who therefore sought to advance his idea for an international Library of Natural Sounds by other means.

The plans for the Animal Phonography Collection and Cornell's own Library of Natural Sounds reveal an ambition to preserve and make accessible the wealth of recorded materials opened up by advancing recording techniques. By the early 1950s, magnetic tape recorders had entered the consumer market, and although these were initially bulky and unreliable, portable battery-powered recorders soon offered a practical alternative to expensive and technically complex gramophone recording equipment. This attracted the interest of zoologists and aspiring bioacousticians, who increasingly regarded animal voices as a key site for studying a host of systematic, ecological, and behavioral problems. By the early 1950s, early adopters of the tape recorder in bioacoustics had already started to make their personal research collections publicly accessible.[4] In the mid-1970s, natural sound archives proliferated internationally and massively; zoologists in the USSR, Europe, North America, and Africa instituted new sound collections for their research, while sound preservationists designated a separate department for environmental sounds in their archives, chief among them the projected British Institute for Recorded Sound.[5]

Among this plethora of initiatives, the Cornell Library of Natural Sounds (hereafter CLNS) and the animal phonography collection envisaged by the International Committee of Bioacoustics were two early, and very divergent, approaches to building repositories of wildlife sound. The ICBA ultimately hoped for a research collection of specimens that encompassed all possible subjects but that would be open exclusively to professional bioacousticians. The expansion and continued existence of the CLNS, in contrast, soon hinged on the involvement of a broad spectrum of collaborators. Kellogg tried to inspire the enthusiasm of academic biologists and aspiring bioacousticians, but he also built extensive and productive connections with commercial publishers, movie corporations, and public broadcasters. And he enlisted a growing group of amateur ornithologists, technology enthusiasts, and "sound hunters" to help him.[6] Doing so required a flexible understanding of the exchange value of animal sound recordings—not just as data samples, but as assets and trophies too. The resulting exchanges between scientific, commercial, and volunteer contexts were what made it possible to muster together two vital resources that the ICBA lacked: adequate funding and a steady supply of samples. In examining how the CLNS expanded into one of the world's largest repositories of animal sound recordings, this chapter brings into focus the importance of exchange to the emergence and consolidation of scientific collaborations. As such, it raises questions about how the knowledge commons could or should be governed.

How Exchanges Are Built

Recent scholarship on data-collecting practices, particularly in data-intensive field sciences such as ecology and biodiversity research, shows that attempts to pool scientific data have often been thwarted by constraints such as the high cost of maintenance, a lack of incentives to share, distrust between participants, and incongruous interests, standards, technologies, or even ways of knowing.[7] Such incongruities are often amplified in collaborations between professional academics and communities of volunteer, amateur, or citizen scientists (Ellis and Waterton 2004; Lawrence and Turnhout 2010).

For the CLNS's reliance on academic, commercial, and amateur contexts to work as it did, therefore, these contexts' distinct regimes of data production, access, ownership, and credit attribution all had to be coordinated and practices of collecting, storing, and exchange had to be streamlined. Sociologies of science typically frame such challenges in terms of the mechanisms and technologies that allow different knowledge regimes to be coordinated.[8] This is particularly useful in elucidating the mechanisms of translation through which such heterogeneous partners manage to arrive at the minimal commensurability required for them to collaborate across their different domains. It is less effective, however, for analyzing how collaborations consolidate over time. To understand how the CLNS and its various collaborations took the shape that they did, became successful, and survived, we must attend not just to the mutual attunement of discourses, technologies, or worldviews, but also to the social operations that sustain them. Collaborations are driven by willingness, and that willingness needs to be cultivated.

Tracing how the diverse groups involved in the CLNS came to be tied into a committed exchange relation requires taking what science sociologists Stephen Hilgartner and Sherry Brandt-Rauf have termed a "data streams" perspective. Data, Hilgartner and Brandt-Rauf hold, do not exist as finished packages. They take shape in continuously evolving flows: flows of information that are continuously produced, reproduced, and reinterpreted, and flows of other resources, such as the money, people, skills, techniques, and instruments that help to generate and transport information (Hilgartner and Brandt-Rauf 1994, 359). Approaching the CLNS and its history as a composite of data streams calls attention to the multidimensional processes by which varied collaborations are brokered: the success of the CLNS's expansion is inseparable from how recording work is funded, how new material is acquired, how participants are willing and able to work with proper instrumentation, submit to standardized protocols, validate each other's work, and ultimately, manage to convince peers of their reliability. Importantly, "data" does not just stream from collaborators to the archive, but enters into a cycle of investment. By situating the library at the interfaces between academic, commercial, and amateur contexts, Kellogg initiated a creative and strategic economy of exchange using the limited resources at the library's disposal—specifically, its collection of sound recordings.

Exchanging recordings enabled each group to barter mutual benefits. In making visible these microeconomics, I draw on a set of analytic distinctions derived from social capital theory (Bourdieu 1986; Coleman 1988). They show how researchers and recordists not only managed to make their recordings productive as economic capital, through stipulations of ownership and copyright, but also converted them into social and symbolic capital within their respective fields. Building on the sociological premise that exchange relations enable actors to develop durable bonds with mutual trust and regard, I argue that the exchange of recordings in a range of socially productive local currencies was instrumental in fostering a "moral economy" that in turn helped the CLNS expand as a scientific archive (see especially Homans 1958). With roots in social and economic history, the notion of moral economy has been variously appropriated by students of modern science to describe the interplay of implicit norms, values, and conventions by which local communities operate (Daston 1995; McCray 2000; Rasmussen 2004; Shapin 1994). It is, in Robert Kohler's words (1994, 12, 253), a piece of social machinery, consisting of unstated moral rules that define mutual expectations and obligations within a community of collaborators, from access to tools to the distribution of credit and rewards. With no shared identity or purpose to fall back on, it is through strategic brokerage of recordings that individual field recordists subscribed to a set of communal standards and values.

Contexts of Field Recording

Broadcasting, Entertainment, and Recording Industries

As chapter 3 has shown, the popular appeal and commercial application of birdsong recordings—as movie soundtracks, gramophone records, radio broadcasts, or soundbooks—was what prompted the production of the first wildlife sound recordings. These applications would continue to shape biologists' use of animal recordings, not just for their own scientific research but also for purposes of education, popularization, and entertainment.

During the roughly ten years that Albert Brand had spent recording birds at Cornell, he released at least six records—some through publishers such as Victor, but most by his own private means.[9] These recordings proved immensely popular, and after Brand's death in 1940 Kellogg and Allen continued to publish new gramophone records under the auspices of their Albert Brand Bird Song Foundation. By 1960, they had published another dozen gramophone records, on their own and with various publishing houses. With titles such as *American Bird Songs* (1950), *Songbirds of America in Color, Sound and Story* (1954), and *Dawn in a Duckblind* (1960), the Laboratory of Ornithology's portfolio contained a mix of educational and entertainment features for a broad audience.[10] Some of these records closely followed the structure of the ornithological field guide: *A Field Guide to Western Bird Songs* (1962c), for instance, accompanied Peterson's *Field Guide to the Birds*, with a human narrator announcing brief excerpts of typical songs to aid field identification (see plate 3). But the Cornell ornithologists also developed far more engaging formats. In *An Evening at Sapsucker Woods* (1958), narrator Arthur Allen personally guided listeners through the woods surrounding the Cornell Laboratory of Ornithology buildings, following the evening chorus of local birds. Other publications were less closely tailored to naturalist purposes. On the record *Symphony of the Birds*

(1960), for example, Kellogg collaborated with Columbia Broadcasting System engineer James Fassett to piece together a selection of Cornell birdsong recordings, accelerating or slowing them down to form an artistic composition that featured as a special intermission in a CBS broadcast of the New York Philharmonic symphony orchestra. Kellogg, Allen, and their graduate students also hosted a weekly half hour on regional radio, where they played their own recordings and discussed local ornithology (Little 2003).

Just as academic ornithologists continued to seek outlets in publishing and broadcasting, so members of the recording and broadcasting industry kept close company with naturalists and students of birdsong. Typically, they united their professional expertise in recording with an amateur interest in ornithology.[11] The BBC, for instance, relied on professional naturalists to host its natural history programming immediately after the war. Chief among these were the conservationist and ornithologist Peter Scott, founder of the research station at the Severn Wildfowl Trust, and James Fisher, the author of bestselling popularizing books on birdwatching and natural history. However, no one embodied the mutual interests of entertainment and scientific naturalism more than Ludwig Koch, whom we encountered in chapter 3 and whose personal and professional connections with the BBC anchored his recording work firmly at the intersection of these domains.

Early in the 1940s, Koch worked as a wildlife recordist for the British Broadcasting Corporation. Vocalizations being practically the only means to capture the radio public's imagination of animals otherwise invisible to them, Koch's collection of wildlife recordings were essential for popular long-running radio and television series such as *Birds in Britain* and *The Naturalist*.[12] Its expanding natural history broadcasts persuaded the BBC to acquire the collection in 1948 (by then, it had grown to include several thousand cuts), and Koch was employed to assist in cataloging it into the BBC's own sound library until his retirement in 1951. As a publicly funded organization, the BBC was in a unique position to maintain and exploit the library as a collection of natural history sounds, not just for program use but for scientific purposes (Burton 1967, 109).[13] With Julian Huxley as its first director-general, the United Nations Educational, Scientific and Cultural Organization (UNESCO) had passed a resolution stipulating the preservation of Koch's collection as cultural heritage, and E. Max Nicholson, whose British Trust for Ornithology lacked the funds to take on the collection, managed to convince the BBC's director-general William Haley to take stewardship. Koch and his patrons insisted that the collection also be made available for scientific use, and in 1953 the BBC set up an informal panel to advise on its deployment, composed of BBC curators, influential public scientists such as Julian Huxley and E. Max Nicholson, the Cambridge University birdsong biologist William H. Thorpe, and representatives of various naturalist organizations (such as the Severn Wildfowl Trust, the Royal Society for the Protection of Birds, and the BTO). In the words of the panel's chairman, Max Nicholson, it was to "form a *link* and a *sieve* between those interested in the scientific use of the collection and the BBC which took care of its preservation."[14] The advisory panel established connections with both amateurs and professionals and engaged them in bioacoustic research on the collection.[15]

Formally, the panel had only an advisory function, yet particularly in the first years of its existence it successfully followed its own agenda, working to legitimate acquisition and distribution policies that were oriented as much toward naturalist aims of preservation and scientific use as toward the BBC's own broadcasting needs. An amateur ornithologist himself, the BBC's newly appointed wildlife librarian Eric Simms actively pursued strong relationships with academic and amateur ornithologists, in part by presenting his own recording work on behalf of the BBC at academic gatherings (Simms 1976, 88). At the International Ornithological Congress in Switzerland in 1954, for instance, the doyen of German ornithology, Erwin Stresemann, offered to extend his network to help the BBC where possible—to locate species or populations of interest to broadcasting and science, solicit interesting records from amateur recordists, or find local assistance in organizing recording expeditions.[16] Simms did not shy away from recording techniques that were particularly helpful for scientists but that had not been approved by his predecessor Koch, such as the parabolic microphone reflector, which was generally associated with an aesthetic that privileged the individual species over its environment. Scientific members of the panel, meanwhile, often suggested avenues for expanding the collection that were in keeping with their own interests. In 1957, the collection was assigned to the newly established Natural History Unit at the BBC, where it continued to develop through acquisitions from amateur recordists and other broadcasting institutions in Britain and abroad, although from 1962 the new curator, John Burton, overturned Simms's objective by shifting the unit's emphasis more to "programme needs than to ... zoological interest," thus adding substantially more habitat or "atmosphere" recordings than the species-specific vocalizations preferred by scientists (Burton 1967, 109) (see plate 4).

The BBC's initial commitment to collecting records of scientific value made it a relevant actor in the expansion of the CLNS, and I trace some of the exchanges between the two institutions in the sections that follow. But the BBC also shaped the network of field sound recordists in other, less direct, but nonetheless unmistakable ways. Its work in broadcasting and discussing nature recordings on air helped to enhance the public's perception of birds as sounding species. By the 1950s, science popularization on the radio was a tried-and-tested format, and the plentiful availability of sound material made birdsong a popular topic. Although not all listeners would have been equally attentive, for some these series would have instilled an appreciation or enthusiasm for, even a passive knowledge of, the songs of common British bird species.[17] Furthermore, scholars such as the behaviorist Peter Marler or the philosopher Charles Hartshorne appeared on BBC radio to explain the science of birdsong. These appearances were an opportunity to increase understanding of their field and alert the general public to the positive implications of more systematic attention to birdsong.

Through broadcasts and gramophone records, then, both the BBC and the Cornell lab addressed large groups of interested listeners. Indirectly yet effectively, they connected the producers of birdsong recordings to a diverse audience of citizens that included a growing group of amateur naturalists with keen ears and a desire to produce their own recordings. For these aspiring recordists, the recordings also established a benchmark for their own work. By selecting what were considered "typical" and the "technically best" excerpts, radio and gramophone records standardized a discourse on the accepted behavior of each species as well as the accepted norms of recording.

Sound Hunters

In the late 1940s, the commercialization of portable magnetic tape recordists appealed to a new group of private individuals with an interest in radio, electronics, and later hi-fi culture. Drawing on a professional background or amateur interest in radio or electrical engineering, such recordists—typically white, middle-aged, affluent men, not infrequently accompanied by their wives—assembled their equipment from a kit of parts, or acquired their own device when the first moderately priced portable tape recorders reached the market. By 1950, amateur recordists had begun to look beyond the consumer applications suggested by manufacturers, such as recording the voices of their acquaintances or copying programs from their radio receiver to play back at leisure (Bijsterveld 2004). Whether these recordists were returning to a long-held interest in naturalist study or saw a fresh technical challenge in outdoor recording, they fashioned their experiments with the air of a new pursuit: hunting the sounds of nature with a microphone. Although early in the 1950s some tape correspondence clubs had been formed, nature recordists were not usually affiliated with wildlife recording societies until the late 1960s (Bijsterveld 2004, 616). Many, however, quickly found their way to prominent producers of nature sound recordings, such as the Cornell Laboratory of Ornithology or the BBC. Because of the publicity these organizations had managed to generate, they were highly visible and widely renowned as experts in field recording. Due to their interest in hi-fi culture, many sound hunters regarded recognition by these recording organizations as an emblem of their own technical craftsmanship. They were often keen for advice or opportunities to collaborate.

Recording organizations welcomed the sound hunters. The Cornell ornithology staff, in particular, maintained good relations with dozens of nature recordists.[18] As one of the driving forces of the CLNS in the 1950s and 1960s, Kellogg was happy to offer "leadership, inspiration, encouragement, material help, and recognition to anyone wishing to study living birds."[19] Kellogg's close corporate links to the Amplifier Corporation of America, for instance, allowed him to tailor the design of its Magnemite tape recorder, the first lightweight and portable model, to fit the needs and financial means of field recordists such as the amateurs he collaborated with. Customized designs like that of the Magnemite gave field recording its own corporate niche and blurred the boundary between manufacturers and users, commercializing technical modifications in an affordable, robust instrument (see plate 5). Along with adjustments to increase the recorder's reliability under the adverse conditions of the field (such as better stability in recording speed), one crucial addition was a functionality that allowed a recorded signal to be monitored while recording was still underway. This greatly increased the efficiency and quality of a recordist's efforts, because it eliminated the need for extensive relistening (Little 2003, 87). Moreover, Kellogg invested considerable time in correspondence and informal workshops with amateur recordists, advising them on how to assemble a suitable and affordable set of equipment or how to improve their recordings to reach "professional" standards, while also calibrating their devices and encouraging them to produce recordings of scientific value.[20] Starting from 1960, the Lab of Ornithology also sponsored a quarterly *Bioacoustics Bulletin*, which institutionalized such technical guidance to aspiring field recordists.

If technical advice was easy to provide, however, technical craftsman-ship alone did not suffice for the lab's purposes. For recordings to be scientifically worthwhile, they had to be free of static, distortion, and ambient noises, provide correct identifications, and include as much naturalistic observation as possible. Despite these elaborate requirements, a considerable number of recordists developed a more long-term and often formal attachment to the Cornell lab, leaving their mark on its Library of Natural Sounds and its bioacoustics research. Indeed, the lab depended on contributions by individual recordists for much of the material that entered its library in the next two decades. In the mid-1950s, it engaged more than a dozen freelance recordists on nearly as many long-running projects. In addition, the CLNS benefited from substantial donations by several dozen individuals.

These amateur recordists included the retired businessman Jerry Stillwell and his wife Norma, who sold their belongings and traveled the United States in a trailer to collect records of American bird species (see figure 4.1). Dr. George Reynard, a geneticist by profession, focused his study and recording work on species on the American East Coast and in the Caribbean. Retired civil engineer Irby Davis specialized in Mexican species and taxonomic relations. Colonel Donald McChesney and his wife Marian specialized in ducks and geese and led recording expeditions to Florida, Kenya, and Canada, while chemist Paul Schwartz established a satellite library of natural sounds for South America in Venezuela. The British ornithologist Myles E. North studied and recorded African bird vocalizations during his tenure and after his retirement as a district commissioner in Kenya (Kellogg 1962b). Their specializations in the field occasionally led these sound hunters to collaborate with full-time academic biologists on publications or conference presentations. It was the record collections, however, that were most often recognized as their major ornithological achievements (see, for instance, Thorpe 1968).

Bioacousticians

After several years of testing, in 1949 the Cornell recordists abandoned their cumbersome disc cutters and sound-film recorders in favor of portable magnetic recorders. It took the academic community a few more years to grow attentive to the new analytic possibilities afforded by these technical innovations. When the twelve recordists that Kellogg brought together in 1951 at a small symposium during the American Ornithologists' Union annual meeting played their recordings, they were more concerned with the technical difficulties of field recording than with its analytic implications.

Figure 4.1
Amateur recording couple Jerry and Norma Stillwell with their equipment in the field.
Source: N. Stillwell, *Bird Song: Adventures and Techniques in Recording the Songs of American Birds* (New York: Doubleday, 1964), 51.

Amateur sound hunters Stillwell, North, and Davis were joined by professional scientists such as the Ohio entomologist Donald Borror, who had started taping birdsongs to teach his ornithology classes, or William R. Fish, who, as a chemist at the Naval Ordnance test station at China Lake, had been able to access commercial equipment to analyze his amateur recordings of California's Bewick's wren. Fish was one of the only presenters to argue that this combination of recording and analysis held "great promise as tools not only for the taxonomists but also for those engaged in studies of avian life histories."[21]

When Fish organized a follow-up symposium in 1956 (the same year that the International Committee of Bioacoustics was established), the focus of the program had clearly shifted. The participants' interest in recording birdsong had become professional and their approach systematic. Ohio ornithologist Wesley Lanyon, for instance, presented his research detailing the differences between the songs of eastern and western meadowlarks. Kellogg's graduate student Robert Stein discussed the role of song as an isolating mechanism between different species in the genus *Empidonax*; his Cornell colleague William C. Dilger offered a behavioral study of how thrush species recognize each other by voice (Mayfield 1957, 86). In barely five years, the ornithologists had transformed the tape recorder from a pleasurable pastime into a way of generating meaningful data about more than the life histories of individual birds—by turning their microphones to particular species in specific locations, they appropriated sound recording as a tool in the study of topics of central concern to biology, such as speciation and variation.

This turn to the biological implications of birdsong studies followed a more general shift that had been taking place in ornithology since the 1930s. By the end of the Second World War, ornithology had expanded from a narrow specialization into a broad-based discipline. According to the existing historiography, a small group of young ornithologists transformed the field from a natural history discipline that had traditionally been divided between systematics (taxonomic studies) and field ornithology into a broad biological discipline concerned with questions of evolution and speciation (Haffer 2008; Johnson 2004). Such developments had been leading in the United States and Germany. But looking back in 1959, British ornithologist Reginald Ernest Moreau triumphantly declared a revolutionary change in the preceding two decades, which he saw reflected in the British ornithological journal *Ibis*. British ornithology, he argued, no longer seemed "inbred and isolated from the main currents of biological science," but instead followed a new interest in "biological papers," including work on migration, territory, population dynamics, behavior, and new systematics (Moreau 1959, 29; see also Johnson 2004).

Studies of birdsong had in fact lagged behind in that development, for although field ornithologists had observed for decades how birdsong varied individually and geographically, little had been done to conserve that diversity sonically. The initial cost and technical complexity of electrical recording outfits meant that the dispersed efforts of pioneering field recordists had been directed at assembling representative records for as many species as possible, rather than documenting subtle variations within single species or even populations. As a portable and affordable alternative to the disc cutter, with tape cheap enough to run at a stretch, the magnetic tape recorder enabled ornithologists to collect recordings in geographically dispersed areas or in successive seasons as individuals practiced and developed their songs. As the next chapter discusses in more detail, another technology also contributed to this development. The sound spectrograph, an instrument that enabled acoustic spectra to be analyzed and visually represented, gave biologists unprecedented control over their recordings and a new way of interpreting subtle acoustic differences as signs of selective variations or innate learning.

Together, the sound spectrograph and the magnetic tape recorder initiated a dramatic postwar boom in studies of animal communication, not just in the United States but in Germany and Britain as well (Marler 2004, 1, 7; Baker 2001, 3). In Germany, Günter Tembrock at the Humboldt University Berlin and Gerhard Thielcke at Freiburg University acquired tape recorders in the mid-1950s, applying them to behavioral studies of chaffinches and tree creepers respectively.[22] In Britain, as we shall see in chapter 5, William H. Thorpe and his doctoral student Peter Marler applied a magnetic tape recorder to the songs of young chaffinches that had been isolated to varying degrees from their parents' tutelage; they concluded that the birds had an inborn template for song learning, which was refined at a later stage through learning by example (Thorpe 1954). Marler used the tape recorder to document variations in the calls of chaffinches around the group's Madingley field station. In a series of papers, he linked the acoustic properties of calls to their specific functions in the social life and habitat of the birds.

Tape recordings were also increasingly deployed in experimental field studies. Researchers had discovered that by playing back prerecorded songs in a given habitat, they could elicit particular behavior, which could then be observed (Niethammer 1955). Population ecologists used this technique to map the extent of a bird's territory by studying the reach of its defensive behavior in response to songs played back from neighboring territories. In 1956, one of these researchers, the Canadian bioacoustician Bruce Falls, played back ovenbird songs to demonstrate that birds are generally capable of discriminating between neighboring birds and visiting strangers. By recording the birds' reactions and editing them into artificial sequences, he also sought to determine the exact song components that birds relied on to recognize each other (Falls 1992, 14). Other researchers used playback experiments as a way of distinguishing the auditory cues used by herring gulls to communicate the availability of food (Frings et al. 1955). These experiments proved useful in designing sonic techniques for dispersing bird flocks, for instance, around airports. Tape recordings also served as ways of monitoring the field during such experiments or for purposes of census and identification. As Hubert Frings and his colleagues pointed out in

1955, humans have a tendency to focus their attention on some aspect of the environment to the exclusion of others, whereas recording the experiment and playing back the tape recording later, "without visual distractions, quickly reveals the true situation" (Frings et al. 1955, 168–169). Tape recorders were used to survey migration patterns of species by means of their vocalizations at night, when visual identification was impossible (Graber and Cochran 1959). These examples illustrate the diversity of projects that began to make use of the magnetic tape recorder in observing, monitoring, and documenting sound.

With Stein and Dilger, the Cornell ornithology group had begun to develop its own line of bioacoustic research, but Kellogg's main concern remained the Library of Natural Sounds. As bioacoustic studies of birdsong took flight in such diverse directions, and the idea of an International Phonography Collection having faded, Kellogg attempted to establish the Library of Natural Sounds more explicitly as a research instrument for the emerging field. This met with resistance, as some professionals remained deeply suspicious of the potential use and reuse of sound recordings. In an application that Kellogg and Stein submitted to the National Science Foundation in 1962 they argued not only that the planned library would be crucial to Cornell's own ethological investigations, but that it would offer long-term intellectual benefits to the study of systematic and ecological problems more generally. Recorded sounds functioned in "parallel to morphological specimens," but also had other advantages: they provided "documentary material about behavioral sequences" and enabled documentation of "the ontogeny of individual sound patterns."[23] In other words, sound recordings complied with a long tradition of specimen collecting, but could as easily be integrated with new approaches in ecology or ethology. The proposal pushed for a more systematic approach to sound collecting, but not everyone saw the purpose of this. One biologist advised the Cornell administrators, "speaking as a systematist," that "the usefulness of your work to other ornithologists may be limited by the amount of data that you can supply with each recording. ... If the bird is unusual, I am sure a systematist would want to have the bird in the hand to go with the voice. At least he would not be satisfied to have one of your foreign collaborators making the identification for him."[24]

Skepticism of this kind prolonged the ongoing struggles for legitimacy of observations between field ornithologists and systematists dating from the beginning of the twentieth century.[25] Systematists had traditionally worked with morphological specimens and had learned to value their aura of material presence; sound recordings forced them instead to rely on auditory and visual identifications made by others. But this skepticism also illustrates a more general problem that the CLNS faced. The tape recorder had greatly increased the ease with which field researchers—both amateur and professional—could independently collect data, and as a result larger bodies of recordings became available. If such recordings were not just to document a species' repertoire of vocalizations but also to provide insight into the geographic variations, taxonomic distinctions, processes of evolution, or behavioral functions that ornithologists were increasingly becoming interested in across a range of different methodological specializations, they had to be reconstituted as "data." And for such specimens to become authoritative in this way, these items had to be produced, edited, and annotated in specific ways.

Trading Copyright

The CLNS managed to consolidate its sound archive as a scientific instrument by regarding it less as an archive of dormant specimens than as a collection of assets. As a scientific archive, the CLNS provided recordings on request to dozens of European and American laboratories in fundamental and applied research (such as the Cambridge University zoology department or MIT's Electronics Research Lab). For purposes that were demonstrably related to education, professional research, conservation, or government projects, the library made its material available at nominal cost, charging only for the cost of tape and the hours that staff spent copying. But the library and its collaborators soon learned to regard the recordings as an economic resource. Recordings could be made profitable as economic capital: copyrighted material could be traded for money, instruments, or other resources that could help to sustain a constant supply of sound recordings.

As a copyright owner, the Laboratory of Ornithology profited not only from royalties on the publication of its recordings as soundbooks and gramophone albums, but also from commercially licensing items in their extensive catalog of copyrighted recordings.[26] The library accepted, for instance, requests from specialized businesses (such as pest control, which used recordings of natural predators to scare off vermin) or the entertainment industry (such as Disney or Warner Brothers, which used recordings for sound effects in film, radio, or advertising).[27] Typically charging a royalty fee of a hundred dollars per species, by 1958 the CLNS had managed to generate a profit of more than ten thousand dollars.[28] Such income from the commercial exploitation of its portfolio, which Kellogg and Allen had transferred to the Cornell Trust for Ornithology, covered a substantial part of the Laboratory of Ornithology's annual expenses. This income accounted for at least half of the budget for supplies, materials, salaries, and benefits in the first few years, the other half coming from gifts to the laboratory.[29] The income the CLNS generated helped convince the Cornell University administration that the Laboratory of Ornithology could be financially self-sustainable, an important factor in its recognition as a research institute, independent of the Department of Conservation. Revolving around Kellogg and Allen and their graduate assistants, the lab had existed only in name (Little 2003, 98). But in 1955, it was placed directly under the president and governed by a specially appointed council. As a self-supporting institute, however, the laboratory was also forced to cultivate additional sources of income, which it derived among others from membership and copyright fees.

The Cornell Laboratory of Ornithology carved out a position for itself at the boundary between public scientific archive and semicommercial organization. This was a tricky position to sustain. On the one hand, the lab was careful to maintain a clearly academic status: the university was entitled to tax exemptions as an educational institution, and this prohibited the lab from engaging too openly in commercial activities.[30] As a result, it had to abstain from advertisements in national media or long-term cooperation with public broadcasters. Most of its recordings were published either through Cornell University Press or with its own funding—although these tactics changed in 1961, when the group assigned all marketing rights to the Houghton Mifflin publishing company, whose promotional efforts increased sales considerably (Little 2003, 196).

On the other hand, the boundary status of the CLNS was not always well understood by its users. Several scientific and nonprofit organizations, for example, repurposed recordings they had received free of charge for scientific purposes to include in popularizing lectures, often for a paying audience.[31] This outraged Kellogg and his colleagues: they considered the CLNS's record publications a valuable asset in securing public support for its own scientific activities, but saw all unauthorized uses of its recordings as piracy.[32] In treating its recordings not just as scientific data but also as assets, the Cornell laboratory sought to restrict the permitted scientific uses of recordings to research alone—much like the BBC, which had made an explicit commitment to the scientific exploitation of its archives by several British naturalist organizations, but did not tolerate its "broadcasting assets" being used in any commercial or popularizing scientific activities.[33] One of the measures taken by the CLNS to curb the unsanctioned use of its recordings was the requirement of approval by a head of staff at a research institute. According to Kellogg, the reason was simply that the archive lacked the time and resources to accommodate elaborate requests. But it was also the case that CLNS recordings were increasingly proving to be the valuable foundation of an exchange economy by means of which Kellogg could maintain the lab's activities.

Copyright granted its possessors an economic resource that could be made profitable in at least two ways. First, recordings could be traded directly with other natural sound archives. A recording's use could be legally restricted, offering a bargaining chip with which to co-opt exchange partners. In the early 1950s, for instance, Kellogg tried to persuade the BBC to donate its recordings to the Library of Natural Sounds, invoking their shared scientific ambitions and the BBC's generosity in distributing recordings among British scientific institutions.[34] The BBC did not accede to Kellogg's plea, and insisted on trading its recordings instead. The Cornell lab finally agreed to the deal, allowing a selection of BBC recordings to be deposited for exclusively scientific use at the CLNS in exchange for a commensurate selection of Cornell recordings and the suspension of their copyright for BBC public service broadcasts.[35] Negotiating the agreement, the CLNS and the BBC anticipated that, apart from enlarging their own sound archives, it would also yield secondary, indirect advantages. For example, the CLNS hoped to reach out to an emerging group of British and continental bioacousticians who might peruse its archives.[36] The BBC, in turn, obliged Cornell to refer field recordists from areas underrepresented in its own sound archive to BBC services, giving it the opportunity to purchase desirable recordings directly from the recordists.[37] In both cases, accommodating each other's recordings was expected to generate visibility and useful direct exchanges with new field collaborators in the future.

Second, copyright provided the CLNS with a steady income that could be reinvested in its scientific operations and efforts to collect more data with the aid of a whole troupe of amateur recordists and beginning bioacousticians. The income was used to acquire professional recording equipment to be loaned to field recordists, or to arrange for their transport or accommodation in the field. Often, such material support was made available explicitly as an investment, on the stated condition that all materials (and their copyright) resulting from the expedition would become the property of the lab—although they would also remain available to the field recordist.[38] In other cases, rather than offering a small stipend to cover the recordists' field costs, the CLNS proposed to purchase their recordings and copyright directly from them as a form of financial support:[39]

We both regard the analytical studies which you are making of birdsongs as very important contributions and if a thousand dollars would help you with your work, and in clearing up some of your equipment problems, I believe the group here could be convinced rather easily to purchase the rights for publication of this material, even without assurance that they would get their money back. If it did come back, then there would be more available for further publication or grants.[40]

Such arrangements, with a scientific institution providing financial and material support to private, often amateur individuals, are rather unconventional in the history of amateur-professional relations, where amateur scientists are usually expected to contribute out of love for science. Yet this kind of patronage, and particularly the exchange of copyright for goods, was advantageous to both the library and its contributors. It brought recordings within the authority of the library; as copyright owner, the institution would be entitled either to distribute the recordings freely for the benefit of the bioacoustics community or to capitalize on them by exchanging them, publishing them, or reselling them. Although no financial return on individual recordings was guaranteed, as a whole such arrangements suggest a cyclical stream of reinvestment in data and copyright that enabled the further expansion of the collection. In turn, transactions of this kind rarely earned amateur recordists a living, but did offer them a way to fund their hobby. Field recordists were often well aware of the desirability and possible capital value of their recordings. Some sold their recordings to peers in ornithological journals or to the BBC, while others offered their recordings in preparing for an expedition in exchange for material, logistical, or administrative support from the Cornell laboratory or individual bioacousticians.[41]

The CLNS was not always in a position simply to acquire the copyright to a record collection, however, and sometimes gave private recordists the option of retaining their copyright while accommodating their recordings in the archive. This served scientific purposes and was beneficial for the sound hunters, as the CLNS's institutional visibility potentially exposed their recordings to commercial interests and enabled the archive to negotiate on their behalf. Although the financial yield for either party would be slim, for the recordist such prospects were attractive, representing a valuable validation of their technical skills and technique and hence symbolic capital in their community.[42] More importantly, these arrangements gave recordists an incentive not to hoard their collections privately in order to capitalize on them in other ways. In some cases, Kellogg also helped to achieve the ultimate aspiration of many long-standing contributors to the library: publishing a commercial gramophone record of their own recordings. Providing both technical guidance and professional contacts, Kellogg helped the Stillwells, William Fish, L. Irby Davis, George Reynard, and Paul Schwartz to publish their work, with others benefiting directly from the Library's support in obtaining publication contracts.[43]

Applying a purposefully commercial logic drawn from the recording and broadcasting industries, then, the CLNS used copyright to demarcate a dual ownership of their recordings, simultaneously private and public. It established itself not only as an agent of liaison, promoting exchanges between users and producers of sound recordings, but, like the BBC's Bird Song Panel, also as a "sieve" that strategically controlled and limited the recordings' circulation. This enabled the CLNS to make its recordings available for scientific research while still protecting its own commercial interests and those of its contributors. The use of recordings simultaneously as data and as commodity granted the field recordists and the sound archive financial resources that helped to sustain collecting activities. At the same time, the exchange of recordings as economic capital produced more than immediate, calculated effects—recordings in exchange for recordings, copyright for equipment. Because recordings were expected to yield increased visibility and individual recognition, they also generated a much less tangible social and symbolic capital.

Exchanging Favors

In an instrumental reading, the notion that sociologists of collective action have termed "social capital" (Bourdieu 1986; Coleman 1988; Homans 1974) captures the productive value of a durable network of social ties, for instance, through the exchange of favors and social obligations that may be accrued in a group (Lesser 2000). Gestures of material, financial, or social support for amateur recordists constituted a subtle indebtedness toward the donor, the library. Unlike in a purely economic debt, the obligation to reciprocate was implied but not usually specified within a particular currency or time frame. Nevertheless, social capital could become productive for the library by persuading recordists to adopt community norms regarding ownership and access to their recordings, implement standard protocols and aesthetic guidelines in their field practice, and place their expertise in the service of the scientific process.

Loyalty and Reciprocity

One of the long-term effects of the CLNS's generous support policy was that it enabled the library to request contributors' nonmaterial support in return. This could be beneficial when, for instance, the CLNS selected material from its archives for promotional or commercial publication. In such cases, contributing recordists and other copyright holders were asked for permission, but also asked to waive their rights and regard their royalties as a charitable contribution to the library and its scientific activities.[44] As such, the library both partook in and exempted itself from customary copyright practice. Reaffirming its academic status alongside its commercial practice, it aligned itself with the principles of a scientific gift economy (Hagstrom 1982). This two-tier structure of commercial and academic recording exchange was fully recognized by the contributing field recordists themselves. Recordists would often voluntarily share or surrender their copyright in favor of the library. One of the laboratory's amateur associates carefully distinguished between the different regimes of ownership involved:

> Once anything is lodged with the Laboratory's library, then the Laboratory can let anybody have the use of these results either free or for any sum it cares to charge, and such sums charged are, of course, credited to the Laboratory and not to me. If however anybody applies direct to me for the use of records, then I would supply them as available and make my own charge, but without ever surrendering any copyright.[45]

Another stated that he had "delivered these tapes as a gift to the laboratory. Consequently they, like the other tapes which we have given to the Laboratory, are the property of the Laboratory and we have no jurisdiction over their disposition or use."[46]

Such gifts should be considered in the context of a durable bond between amateur recordists and the CLNS. According to anthropologist Marcel Mauss (2006), instances of gift giving are not singular events, but must be interpreted as part of a cumulative cycle of reciprocity. Gifts create social bonds, Mauss argues, because they entail the implicit obligation to reciprocate in the future. In the CLNS setting, the exact conditions of this gift exchange were rarely made explicit—but by donating recordings and waiving jurisdiction, the recordists often felt they were reciprocating the support they had received from the lab. When volunteer Donald McChesney discovered that a colleague had openly objected to the CLNS's plan to share some of his recordings with the bioacoustics community, fearing that this would jeopardize the priority of his own publications, McChesney acerbically observed that the person involved looks upon [the recordings] as lying within his own private domain. This runs somewhat counter to my understanding of scientific work in general and the work of the Laboratory in particular. I had an idea we were all working in a common effort for the advancement of human knowledge. And quite apart from the ethical aspects, hasn't [he] in large measure been supported by the Laboratory or by agencies made available to him through the Laboratory? It seems to me that this might in all fairness, have some bearing on the scientific use of the products of his efforts.[47]

This remark expresses a moral imperative premised on a belief in fundamental scientific norms such as the communal nature of research materials. It also adduces a moral obligation to reciprocate for the CLNS's financial, technical, or social favors with the same loyalty and generosity. Such reciprocity meant that gifts did not simply redress the balance, but also provided individual recordists with credit that would give them access to their own forms of social capital, such as recognition as technically qualified recordists.

Compatibility

Whereas recordists' adoption of community norms ensured the CLNS's access to recordings, their adoption of certain recording standards was intended to guarantee academic and commercial compatibility. For sound recordings to serve as specimens of acoustic data, bioacousticians had to achieve a particular aesthetic, and they required specified protocols to structure the vital metadata. Like the administration of intellectual property, the production of recordings itself was framed by the CLNS's dual scientific and commercial status, as well as the gestures of material and social support it had provided:

> For the privilege of making copies of any of your Mexican bird songs ... we hereby offer you $500. ... We very much hope that on your future recordings you will announce the bird's name, the date, locality, as well as keep adequate field notes concerning the contents of each reel. It would also be of considerable scientific value if you would make a habit to sound a pitch-pipe, preferably the A note (440cps).[48]

Such protocols were important to ensure the scientific value of each recording included in the depository. A standard tone, for instance, enabled future users to calibrate their playback equipment and detect deviations in recording speed that would otherwise be hard to determine by ear.[49] In addition, recordists were required to file their work using standard "field and editing forms." Such forms were used as classic "boundary objects" in the sense that their format was standardized to allow bioacousticians and amateur recordists to exchange valuable information regarding their recordings and field experiences, but plastic enough to adapt to local needs and levels of expertise. The forms left space, for example, for details on specimen collecting, detailed descriptions of behavior, or the possibility of playback and decoys, which would matter to different specializations to different degrees. But although the value of a recording was enhanced substantially when extensive field notes on circumstances and the presumed biological significance of the recorded behavior were included, amateur recordists could not always be expected to complete the form.[50] At a minimum, library administrators asked that field recordists aim to properly identify a specimen and include details on the date and locality of recording. For most recordists, after all, the administration and editing of their recordings was a tedious and often invisible aspect of their work, which only detracted from time spent in the field. Framing their contributions as part of an ongoing exchange of favors rather than as a singular donation provided the library with a way of insisting on diligent completion of the recording and editing forms.

Aesthetic standards for the recording, similarly, were subject to a negotiation of compliance in exchange for favors. The CLNS staff had a long-standing preference for what they termed "focused" or concentrated sound—a recording with a high signal-to-noise ratio, ideally focused on a single individual and with minimal noise. To achieve that, bioacousticians associated with the CLNS promoted the use of a parabolic microphone, a large spherical reflector that concentrated sounds at a microphone in its focal point, dramatically increasing sensitivity and narrowing recording range. This tool was not only efficient and useful when recording birds in flight or at a distance in the field: it also had distinct scientific and commercial advantages for the CLNS. Scientifically, focused recordings tended to produce "cleaner," less noisy spectrograms than traditional recording techniques.[51] And commercially, recordings with a high signal-to-noise ratio were especially marketable, as they had several potential applications. In their raw form, such close-ups served well as sound effects or as samples for ornithological identification guides, but they could also be processed to meet a demand for atmospheric scenes on radio, film, or popular records. As sound walks *avant la lettre*, several of Cornell's own popular gramophone publications, such as *Dawn in a Duckblind* or *An Evening at Sapsucker Woods*, had been composed using individual species recordings.

Recordings by beginning sound hunters, however, were often "out of focus"—oriented toward a habitat rather than an individual bird. Keen to convince field recordists to produce technically outstanding yet focused recordings, Kellogg often cited the standards of production for commercial actors such as Disney or Warner Brothers.[52] But while such technical pride was advantageous commercially, it could be problematic scientifically. To make a recording sound more attractive, it was tempting to intervene by filtering out certain frequencies, editing silent segments from the tape, or simply withholding technically imperfect recordings. For the bioacoustician, that selectivity could destroy potentially valuable information.[53] The CLNS's policy was thus to store all recordings that reached a reasonable quality, even if they could never be used commercially, as they might still contain a great deal of information on singing patterns.[54]

Kellogg and the CLNS explicitly demanded restraint from their field collaborators, who were to abstain from any "aestheticizing" interventions. One recordist was prompted to confirm, for instance, that "as requested, I shall not use the band-pass filter at all in the copying, so you will get the true record including all the hums and buzzes and external sounds which you can of course filter out yourself."[55] Clumsy editing was generally not hard to detect for professional listeners at the Cornell lab and was met with expressions of concern: "It is my feeling from listening to the record that someone has already seriously attenuated the high frequencies and I know that many of the low frequencies have been attenuated. What I would like to do, and I think you should do, is try to get recordings which will need no filtering at all."[56]

Such aspirations regarding the standards of annotation, technical quality, or authenticity in amateur recordings illustrate the social rapport that had to exist between recordists and the CLNS bioacousticians. The CLNS would formulate a set of norms that increased the scientific (and commercial) value of a recording but also placed a burden on the recordists, who were asked to dedicate a degree of time, care, or self-restraint that did not always seem to be in their own immediate interest. However, these "expenditures" represented currency in an ongoing, continuously evolving exchange, in which favors owed could be paid back but also generated new ones.

Collective Listening

Finally, social capital enabled the CLNS to check and improve the quality of data it collected. Recordists were expected to not purposely misreport their data. The exchange of social capital also guaranteed the institution access to a network of knowledgeable reviewers for the recordings. Unlike other types of data that were directly codified and inscribed in the field (such as sightings in ornithology), recordings could be played back and compared infinitely. As such, they remained open to redeliberation or annotation by other listeners—especially important in cases where visual confirmation was lacking. Recordists who contributed a large number of recordings (and most did) would be invited to copy and edit their collection at the CLNS, where they could process their recordings using professional equipment and in collaboration with library staff, who assessed the quality of recordings and assisted in transcribing field notes, identifying additional background sounds, or even reconsidering particular entries to ensure the quality of specimens entering the archive.[57] Students, faculty, and amateur or academic visitors would drop by, lending their ears and expertise to problematic cases, uncertain identifications, or incomplete metadata.[58] Together, these listeners constituted an informal peer review, whose "collective ears" were thought to ensure validation of the recordings.[59]

A still broader group of scientific listeners were engaged at other stages of the process. Returning from the field, for instance, recordists would often sit down to review their samples and identifications with local experts in the region or experienced visitors to the lab. As the geographic scope of the collection began to outgrow the expertise of researchers on-site, the library solicited the opinions of other experts in the network of institutions and researchers who requested recorded material.[60] Although scientific users could procure their material free of charge, they would occasionally be asked for their expertise as a token of exchange in the service of the library, by sifting out suspicious identifications or adding important biological information that could be gathered independently of the recording context.[61] For potentially important yet questionable identifications of a record, the CLNS used its broadcasting and publishing networks to reach out to a larger community of amateur and professional ornithologists. This, one collaborator later recalled, "did a lot to build confidence in the trustworthiness of the LNS as well as to increase interest in improving and expanding the archive even further."[62]

The exchange of social capital helped to sustain relations between the CLNS and its scientific and amateur collaborators, as its reciprocity suggested a basic level of interpersonal trust. By its very nature, the mobilization of social capital is not without risk: because its value is mostly implicit and without the assurance of a binding agreement, participants might choose not to reciprocate at all—instead holding on to their recordings or refusing to adopt standardized protocols. By reciprocating, in other words, collaborators signal their good intentions and establish bonds of solidarity (Homans 1974). Such bonds form a baseline of social emotions that are critical especially in interdisciplinary or heterogeneous collaborations, where collaborators may be dependent on each other's expertise, reliability, and goodwill (Knorr Cetina 1999; Parker and Hackett 2014). In the context of collaboration between the CLNS and field recordists, these reciprocal exchanges were effective not only because they balanced each other out, but also because they could set in motion an "escalating reciprocity" (Farrell 2003), a feeling of mutual obligation that motivates collaborators to match or exceed the standards or quality of their partners' work. As field recordists began contributing larger portions of their collections to the sound archive, the CLNS reciprocated in new ways, by assisting recordists in publishing their own long-playing records or setting up their own spin-off libraries, or by awarding them more symbolic capital. Such escalations may well account for the gradual tightening of collaborative bonds that occurred in the course of the 1950s.

Matters of Distinction

In addition to the economic and social capital that it exchanged, the library introduced several forms of symbolic capital—prestige or recognition—as an effective way of rewarding recordists for their sustained commitment. It did so, first, by furnishing CLNS field collaborators with an identity and a symbolic distinction among recordists. On several occasions, Kellogg pointed out that good birdsong recordists possessed a superior understanding of high fidelity compared with average listeners or even professional commercial recordists, whose ears "are just not trained for such sounds. They tend to tolerate distortions, which are totally unacceptable to a person familiar with the birds in the field."[63] Tapping into many field recordists' aspirations to high-fidelity recording, such judgments sought to relocate "professional audition" into the realm of academic recording and to acknowledge the naturalist-recordist's heightened sensory skill (Porcello 2004, 734).

Other forms of symbolic capital were less dependent on recordists' standing among their nonscientist peers than on recognition within the bioacoustics community itself. The CLNS was keen, for example, to acknowledge recordists and editors as the authors of a recording, by entering their names in the standard editing forms that contained the metadata for each recording. To be sure, authorship of a recording was regarded as distinct from ownership, as described above. As recordings changed from private property to communal resource, recordists could choose to renounce, donate, or sell the copyright for a recording to the CLNS but would not cease to be its author. Recorded data samples required authorship in order to exist within the sound archive, and were not usually accepted or regarded as useful without an explicitly identified author. This was in line with other scientific contexts, where, as Mario Biagioli (1998, 5) has observed, such specification is important "not only because scientists deserve fair credit for [their work], but because it has to be marked in order ... to be recognized as a specific truth, not just a chunk of undifferentiated, undescribed nature."[64]

This notion of authorship was particularly effective because it could reflect both positively and negatively on the individual recordist. On the one hand, authorship of a recording awarded the recordist symbolic credit—as opposed to the economic rewards distributed through copyright. This enabled field recordists to be properly acknowledged in the academic publications, gramophone records, or broadcasts in which their recordings appeared, at no monetary cost to the CLNS. On the other hand, by inscribing the author prominently into the metadata, the recording and its production, identification, or annotation were tied directly to the recordist's reputation. Authorship of a recording, in other words, also assigned accountability. Paired with the cycles of peer feedback described above, whereby different actors—recordists, editors, field experts, bioacousticians—reviewed and sometimes coauthored a recording, this sense of authorship encouraged recordists to consider with the greatest care the information they provided. Badges of responsibility as much as records of recognition, the recording forms affected authors' reputation and standing with the laboratory, and hence their access to other resources. Reputation and trust among library staff could also be converted into more durable forms of symbolic capital, such as the official status of "research associate."

Most of the long-term collaborators and supporters of the Library of Natural Sounds named in this chapter were appointed research associates—an official title that signaled affiliation with the Laboratory of Ornithology but did not require formal qualifications. Research associates were nominated by laboratory staff in recognition of their contributions to the lab and, although they received no salary, they enjoyed full rights as staff members. As Kellogg explained to a nominated associate, this was "of considerable advantage to [the associate] in that it gave him prestige and a good address. The University likewise shared credit for anything he accomplished."[65] The honorary appointment of amateur scientists was not undisputed within the Laboratory of Ornithology. Scientific staff not directly affiliated with the Library of Natural Sounds criticized its liberal policy and felt that research associates should not be appointed without academic qualifications in their field.[66] For CLNS staff, however, the appointment was an important mechanism: at a basic level, it represented the institution's gratitude for the recordists' services, fieldwork, or donations, but the quasi-scientific identity of "research associate" itself was a token of exchange that could perpetuate the mutual commitment and social obligations between the fieldworker and the institution. After all, as staff members, research associates could not only expect continuing academic and material support for their research projects; they were themselves expected to provide continuing support for the advancement of the laboratory's scientific collection by donating their recordings and research papers.

Knowledge Commons

In the two decades after the Second World War, the CLNS transformed from a small collection of Cornell birdsong recordings into an archive housing thousands of records for bioacoustics research. That change required the staff to overcome problems of translation and information management, but also to create the necessary social, material, and financial conditions and to generate engagement and commitment among collaborators. Kellogg and his colleagues at the Cornell Laboratory of Ornithology managed this multidimensional process by crafting an intricate exchange economy. Exchange was premised in part on the principles of an academic gift economy, in which disclosure is traditionally rewarded with academic recognition. But crucially, it was also propelled by trading mechanisms that rewarded disclosure with selective privileges.

All this time, the library had been governed by a moral economy that matched the small and personal scale of its operation. It was not, however, easily adaptable to important changes in its conditions of operation, and in the mid-1960s, its growth gradually faltered. One important change was to the driving forces behind the Library of Natural Sounds: Allen had retired in 1953 and in 1967 Kellogg, too, stepped down from his academic appointment. As a professional bioacoustician, recording pioneer, and avid tinkerer, Kellogg had been key to the library's functioning. A true "boundary shifter," he moved back and forth between academic, commercial, and amateur contexts of field recording, spotting openings and aligning their interests (see Pinch and Trocco 2002). There was no academic succession arranged for Kellogg: the status of separate department in the university that the Laboratory of Ornithology had acquired in 1955 meant it was not entitled to appoint its own faculty. Kellogg remained active as an assistant director of the archive for another six years, but without faculty to assume leadership over the library or graduate students to work on bioacoustics problems, its program came to a grinding halt (Little 2003, 152–153).

During his retirement, Kellogg crucially maintained correspondence with nearly a dozen lab research associates, most of whom still contributed new bird sound recordings. However, they too had reached a considerable age. Moreover, as Michael Harwood (1987, 9) points out, the laboratory's strategic reliance on amateur recordists and commitment to the popularization of the field had put it in an awkward position vis-à-vis the ornithological establishment. The field of ornithology had become increasingly specialized and professionalized, and around this time the respectability of amateur fieldwork experienced a low point. Without its vocal defenders among faculty, amateur engagement cooled for a while. In addition, the university as a whole had been struggling to keep a positive balance sheet in the face of diminishing support from government agencies, corporations, and foundations. Meanwhile, at a time when the laboratory was increasingly expected to sustain itself financially, it experienced great expansion. Although its income from royalties grew steadily, thanks to the continued popularity of its recordings, this could hardly compensate for the pressure of rising operating costs for supplies, building maintenance, salaries, and fringe benefits.

The history of the CLNS offers an interesting counterpoint to a growing literature in sociology, history of science, and management studies on the commercialization of scientific knowledge. Some of these scholars have signaled (and often lamented) an expanding operation of market principles in academia, with corresponding shifts in normative regimes (Etzkowitz 1989; Mirowski and Sent 2002; Nelkin 1984; Zuckerman 1988). Others have shown that the public exchange of ideas and research materials is often strategic and selective, especially in those fields where economic gain can be expected (Evans 2010; McCain 1991). A closer look at the early history of the CLNS shows, however, that rather than sailing by a general Mertonian normative compass (see Merton 1971), local scientific communities often develop their own intricate moral economies, borrowing from their ambient cultures and adjusting to their own social and material conditions. Sound recording was a peculiar kind of scientific material: it could be multiplied without loss, but by convention its authors and owners enjoyed some form of social and legal protection. Within those conditions, an exchange economy emerged in which research materials could be commodified and selectively distributed to reinforce academic interests. As a commercial trope, copyright was not simply applied to turn research materials into private property and economic gain. It served specifically as a strategy to secure financial resources for the CLNS's scientific activities and to enlist other actors in that enterprise while protecting their commercial interests.

At the same time, copyright was imbued with alternative meanings, making it a currency for exchanging both recognition and accountability.

If there is a lesson in the CLNS's exchange economy regarding the governance of the knowledge commons today, it is not necessarily that commercial logics should be applied to modern participatory data-collection ventures. The CLNS case may well provide a historical precedent for the commodification of raw data that has recently been observed in amateur-professional collaborations in biodiversity research (Ellis and Waterton 2005; Lawrence and Turnhout 2010). But its commodification was also embedded in a historically, socially, and materially specific exchange economy. Above all, therefore, this case alerts us to a more general point: the traditional contract of gift exchange that exists between amateur and professional collaborators must not be taken for granted. As much scholarship on amateur, citizen, or volunteer involvement in the field sciences has shown, such naturalists actively bring their own meanings, expectations, and practices to the process of collaboration (Bell et al. 2008; Ellis and Waterton 2004; Lawrence 2006). In view of present-day volunteer communities' discontent with increasingly standardized and rationalized processes in data production, and their application and valuation by distant, unknown users, the CLNS case presents a historical prototype for imagining new variants of productive collaboration between amateur and professional collaborators that include a sensibility for the complexity of data flow processes and potential symmetries in exchange.

5 Patterned Sound: Inscriptions and the Trained Ear in Birdsong Analysis

A Novel Technique

"The invention of a novel analytical technique often helps to launch a new science. What microscopes were for the emergence of cell biology as a discipline, or the cathode-ray oscilloscope for neurophysiology, it was the sound spectrograph that, immediately after the Second World War, enabled the birth of the science of birdsong. … Until about 1950, everyone interested in birdsong had no choice but to work by ear. Only when the sound spectrograph became available was it possible, for the first time, to grapple objectively with the daunting variability of birdsong" (Marler 2004, 2). Thus biologist Peter Marler sketched in a textbook on birdsong biology what he regarded as a formative shift in the discipline.[1] Originating in American telecommunications engineering, the sound spectrograph represented a sound spectrum visually, by plotting frequency in function of time. It was this application that, according to Marler, "elevated studies of song dialects … from the birdwatcher level to the status of scientific research" (Marler 2004, 11). Marler was well placed to comment: together with his superior, William H. Thorpe, at the Cambridge University Department of Zoology, he became a leading figure in a sound visualization technique that would prompt an unprecedented change in the substance and organization of the study of birdsong (figure 5.1).

Among students of bird vocalizations, the sound spectrograph was heralded as the solution to a problem that had preoccupied them since the turn of the century: how to read and represent birdsong objectively and accurately, but also intelligibly. As a device that permitted an analysis of sound in a way that was roughly analogous to human hearing, it became deeply associated with the concept of "visible speech" and the idea that the eye could be trained to read meaningful patterns of sound vocalization in much the same way as the ear could perceive them. For a brief moment, its

Figure 5.1
William H. Thorpe recording a dove at the Madingley Ornithological Field Station
near Cambridge.
Source: J. Hall-Craggs, "Obituary: William Homan Thorpe," *Ibis* 129 (s2) (1987): 568.
Reproduced with kind permission of Les Barden.

developers at the Bell Laboratories considered the instrument particularly
suited to support oral education and visual telephony for the deaf (Mills
2010). But the prospect of a universal language for sounds, based directly on
its physical properties, also found a much broader appeal among students
of birdsong. Ever since Albert Brand (1937b) had critiqued the human ear as
a subjective and unreliable tool of investigation, students of birdsong had
sought to compensate for their own defective hearing with various techniques of sound visualization—albeit with varying success. Yet here was an
instrument that promised to evacuate detailed acoustic analysis of birdsong
from the domain of the audible altogether at no loss, an instrument that
did not merely visualize sound but, so its developers initially projected,
would actually provide an intuitively legible orthography of sound. The
sound spectrograph's promise was to fulfill, as media arts theorist Douglas
Kahn (2002, 180) puts it, a long-standing desire to make sound "tangible
and textual by making the invisible visible and holding the time of sound
still." However appealing, the concept of visible speech proved difficult to

achieve in practice, and communication engineers soon moved on to other applications for the sound spectrograph. But by that time, a new generation of animal behaviorists had adopted the device, along with its developers' initial hopes and assumptions about the transcription and transmission of sound.

It is tempting to read Marler's analogy between the sound spectrograph and visual technologies such as the microscope or oscilloscope as signifying a landslide turn from analysis by ear to an exclusive reliance on visual inscriptions. Indeed, at first glance, these inscriptions may seem to discard the human "expert ear" in favor of a mechanical ear in the recording instrument. In practice, however, the concept of an objective yet intelligible visual language for sound could not be fully sustained. Understanding the ubiquity of sound spectrography in birdsong biology thus requires careful consideration of its embedding in an analytic and representational practice that, although ostensibly visual, was never entirely mute itself. Nor did scientists themselves always want it to be or pretend that it was. In this chapter, I attend to the minutiae of ways of analyzing and representing, listening to, and looking at sound that biologists at the Cambridge University zoology department practiced in their spectrographic work. Tracing the sometimes unsuccessful and controversial ways Cambridge University students of birdsong positioned spectrographic visualization in relation to an embodied experience of sound, this chapter revisits the notion of "inscription" to ask how and why certain visualizations acquired authority in the science of birdsong (Latour and Woolgar 1986).

Visible Speech

The Bell Telephone Laboratories began developing the sound spectrograph in 1941. They initially conceived of it as one of their telephone products, but the project was soon classified because of its potential military relevance. The instrument automatically broke down complex sounds into individual components and represented them in a visual order. This principle found useful applications in wartime cryptanalysis and naval intelligence. It served to expose scrambling in telephone communications and to decode and reconstruct speech fragments, and may also have been used to distinguish friendly and enemy naval craft by their engines' signature sounds (Fehr 2000, 41; Radick 2007, 455). Even before the end of World War II, however, Bell Labs engineer Ralph K. Potter and his colleagues resumed their search for civilian applications, setting up a training program for the deaf to learn to read spectrographically rendered speech (Potter, Kopp, and Green 1947).

The idea of converting speech sounds into a readable script was not new. The phonautograph that Leon-Scott famously developed to elevate stenography to an automatic, universal written language was variously appropriated in the study of human voice.[2] Scott constructed an analog model to the anatomy of the ear and attached to a stylus to render the waveforms produced by speech visible, but the resulting curves were irregular and difficult to decipher (Brain 2015, 74). Nevertheless, modified versions of the device found welcome reception, among others in the mid-1880s by physiologist Victor Hensen, who redevised it into a *Sprachzeichner* or speech depictor, and in 1899 by Yale experimental psychologist Edward Wheeler Scripture, who had set up a program to record speech waves for research on phonetics. Although the curves were about as hard to decipher as "Chinese ideographs," Scripture's hope was that they could ultimately be read in such minute detail as to allow conclusions based purely on what could be observed by the eye. Likewise, French linguists joined forces with experimental physiologist Étienne-Jules Marey to study phonetics by recording air pressure and movements in the nasal passages, larynx, and lips (Brain 2015, 72). In the 1920s, Milton Metfessel—a student of psychologist Carl Seashore at the University of Iowa—developed a technique he called "phonophotography," seeking to record and transcribe singing voices with a detail and accuracy that would allow him to study emotional expression. Each of these developers was motivated by a more or less explicit desire to mute their studies of sound as much as possible. When Potter and his colleagues introduced the sound spectrograph in a 1945 article in *Science* and in the monograph *Visible Speech* in 1947, they placed themselves within this lineage—the term *visible speech* referred to Alexander Melville Bell's system for phonetic transcription.

But Potter's team also explicitly distinguished the innovation from existing methods. The team now argued that existing inscriptions were simply "unreadable to the eye" (Potter, Kopp, and Green 1947, 4)—the problem being not so much the resolution of detail, but the perceptual decoding of the inscriptions. At the time Potter was writing, the preferred instrument for displaying sound was the cathode-ray oscilloscope, which presented variations in acoustic energy over time as a waveform. Such waveforms provided crucial information on the physical nature of sound, but they gave very few clues as to the actual auditory experience. Tellingly, for Potter, if the spectrogram was an outstretched rug with its pattern clearly visible, the waveform was comparable to the rug's threads unraveled and bundled together: the waveform provided abundant information, indeed, but most of it gave very few clues as to the actual perception of sound (Potter, Kopp, and Green 1947, 351).

The sound spectrograph's pattern was produced by etching recorded sounds onto paper in a series of shaded bands, plotted against a horizontal time axis (see figure 5.2). Vertical spacing indicated frequency, while coloring suggested a relative amount of sound energy. According to the sound spectrograph's developers, it produced a "translation [of sound] similar to that made by the ear. It should spread out the dimensions of speech so that they were visible to the eye *as* they are audible to the ear" (Potter, Kopp, and Green 1947, 4). Rather than showing the intricacies of an acoustic waveform, the representation depicted its perception (Mills 2010, 38). The spectrograph enabled people to see what they heard, Potter explained, because it was modeled on a schematic cochlea of the inner ear (Potter, Kopp, and Green 1947, 10–11). The inner ear was believed to be made up of sensitive elements that were each tuned to a particular frequency, the sum of reactions by these elements being what produced a physical sensation of tone. Analogously, the sound spectrograph automatically applied a Fourier analysis by unraveling a complex sound into the simpler sound waves that constituted it. The sound spectrograph thus did for sound what an optical spectrograph could do for white light, by distinguishing its constituent frequencies of color. Put very schematically, the device recorded a sound signal from a magnetic tape recorder, and then looped this signal through a filter that tuned to successive frequency ranges. A stylus then traced the sound energy present in each of these successive frequency bands on a revolving roll of electrically sensitive paper (Koenig, Dunn, and Lacy 1946).

At the beginning of its career, the sound spectrograph—conceived as a mechanical analog to the human ear—had seemed like an ideal substitute for the hard of hearing (Mills 2010). But the genealogy of "visible speech" also reflects a more general conviction that it would be both desirable and possible to replace hearing completely by seeing (Mills 2010; Tillmann 1995). As the history of inscription technologies shows, such instruments were thought of not only as sensory prostheses for the deaf, as a physical necessity, but also as aids to help able-hearing linguists and acousticians transform sound into a scientific object, hence as an epistemic aspiration (Brain 1998; Sterne 2003). These scientists aspired to convert sound into a set of readable signs without the loss of crucial information. Once sound could be "read" from the image, listening, as a private and thus subjective experience, would become public and accessible for all to scrutinize.

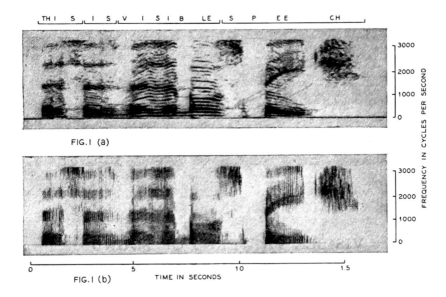

FIG.I (a)

FIG.I (b)   TIME IN SECONDS

Figure 5.2
Original spectrogram of the phrase "This is visible speech." The horizontal axis expresses time, the vertical axis frequency. Amplitude is suggested by the darkness of the trace.
Source: R. K. Potter, "Visible Patterns of Sound," *Science* 102 (2654) (1945): 464. Reproduced with kind permission of AAAS.

The developers of the sound spectrograph assumed that this was possible because both aural and visible languages were made up of patterns, and similar expressions would, they believed, have broadly similar patterns. Their preliminary evidence seemed to suggest that the "indexicality" of sound and image would enable a new kind of visible ABC to be assembled, one that, with sufficient training, anyone could learn to read. With phonetician George Kopp and his psychologist assistant Harriet Green, Potter set up a training program to test this assumption, by training five normally hearing women to learn to understand the visual language (Potter, Kopp, and Green 1947). However, the project failed to yield conclusive proof that visual hearing could be efficiently achieved. Meanwhile, researchers in the Haskins Laboratories tackled the problem from the reverse direction. Physicist Franklin Cooper and psychologist Alvin Liberman set out to design a print-to-sound machine to read out for the blind, but hit problems when their respondents could not learn to "read" the sounds (Fehr 2000; Liberman and Cooper 1972). To discover more about the minimum parameters of intelligible speech perception, they experimented with playing back spectrograms on which they themselves had drawn simple patterns. Although these experiments demonstrated that simple acoustic patterns could actually translate into simple visual patterns (and the reverse), just as the spectrograph developers had anticipated, the displays of actual human speech proved too complex and ambiguous for a human listener to read (Cooper, Liberman, and Borst 1951). It soon became clear that the development of a truly visible language for speech was far from completion, but other applications seemed more feasible. If human speech was too complex for spectrographic reading, simpler patterns often displayed perfectly well. By 1951, Kay Electric Company, a company founded by a former Bell Labs engineer working on US Navy radar control projects during World War II, was licensed to develop the sound spectrograph commercially, naming it the Sona-Graph.[3]

Spectrographic Studies of Bird Communication

The Sona-Graph sound spectrograph was quickly adopted by universities worldwide, and became a preeminent instrument in fields such as linguistic analysis, phonetics, speech analysis, and signals intelligence. It was through exposure to these fields, and often its wartime applications, that students of animal vocalizations first learned about the potential of this new instrument. When Potter (1945, 470) introduced the sound spectrograph in *Science*, he already intimated that biologists might find the instrument useful "to analyze, compare, and classify the songs of birds, and, of even more importance, it will be possible to write about such studies with meaningful sound pictures." He even put his hypothesis to the test. In 1948, he sent a draft paper to Arthur A. Allen at the Cornell Laboratory of Ornithology, documenting some of the findings he had collected by experimenting in his spare time with the spectrographic analysis of frog sounds—which he had taken from the laboratory's gramophone publication *Voices of the Night*.[4] Despite this forewarning, the Cornell ornithologists likely got to know the spectrograph firsthand through Charles Hockett, a linguistic anthropologist in Cornell's Division of Modern Languages who would later pursue studies of animal communication in a bid to define the properties of human language. Likewise, when Cornell zoologist Nicholas Collias published a study on vocalizations of domestic fowl in 1953, he had been provided access to a rewired commercial Kay Sona-Graph by his coauthor, the University of Wisconsin linguist Martin Joos. Joos had worked intensively with the instrument during the war, while employed by the

US Signal Corps to conduct cryptanalysis, and used the spectrograph as the basis for what he termed *acoustic phonetics* (Fehr 2000, 41; Joos 1948). Military service in naval intelligence also likely provided Donald J. Borror, a pioneering bioacoustician, entomologist, and amateur ornithologist, with firsthand experience with the device (Marler and Slabbekoorn 2004, 3). With Carl Reese, a colleague at Ohio State University's Department of Zoology and Entomology, Borror secured access to the instrument in the astronomy department, where it was used to study the scintillations of stars ("Visible Bird Song," 1953).

William H. Thorpe, a Quaker and conscientious objector without military experience during the war, may have first learned about the sound spectrograph through an article by a British engineer and amateur ornithologist in *Ibis*, the journal of the British Ornithologists' Union (Bailey 1950). The author had learned that one of the only spectrographs in Britain was used at the General Post Office Research Station in London and he obtained their consent to use it for serious research requests. Thorpe (1979, 68) had begun his career in insect physiology, but under influence of continental ethologists, particularly the work of Konrad Lorenz, shifted his research toward mechanisms of learning and instinct in animals. Together with Niko Tinbergen, Thorpe was responsible for establishing ethology in England after the war, as editor of the newly establish ethological journal *Behavior* and president of the Association for the Study of Animal Behavior (established in 1936 by Julian Huxley) (Durant 1986). An avid birdwatcher, he decided that birds would be ideal subjects for studying behavior and campaigned for an ornithological field station, which began operations in 1950 on a plot of land in Madingley, near Cambridge (Burkhardt 2005, 341–342). During its first year, Thorpe and the station's newly hired curator Robert Hinde initiated several research projects, the largest of which examined the nature of song learning. Thorpe had already turned to the BBC engineering department and the General Post Office for technical advice on sound equipment, and in 1951 he was able to take a collection of birdsong recordings to the Bell Telephone Laboratories for conversion. There he learned that at least one instrument existed at the Admiralty Research Laboratory, for analyzing submarine noises. Thorpe managed to convince the Admiralty to let him use its spectrograph regularly, until in 1953 he used a Rockefeller Foundation grant to purchase his own for the Madingley field station (Burkhardt 2005, 343). In 1951, Thorpe was joined by several new staff, including botany graduate Peter Marler, who was starting a doctorate on chaffinch behavior. Together with Thorpe, Marler learned to use the sound spectrograph on their own recordings and the copies of BBC record-ings that Thorpe had been given by its Bird Song Panel.

In the sound spectrograph, Thorpe (1954, 465) identified an alternative to existing methods of birdsong analysis, which, he found, all suffered from "the primary difficulty of perceiving accurately by the naked ear elaborate sound patterns of high frequency, high speed and rapid modulation," so that "vocalizations were formerly the most difficult of all [behavioral] releasers to investigate precisely." Thorpe was glad to announce that "they have now become far more readily amenable to analysis than many patterns of visual and olfactory stimulation." Sonagrams—an abbreviation for the Sona-Graph's sound spectrograms—supplied both a new form of notation and a method of precise measurement that allowed analysts to avoid the "dan-gers of subjective interpretation" entailed by earlier notation technologies (Thorpe 1954, 465). Other pioneering spectrographic investigations of bird vocalizations presented similar arguments. Donald Borror and Carl Reese explained that "most accounts were merely subjective descriptions and not accurate analysis" (Borror and Reese 1953, 271). While acknowledging that naturalists like Aretas Saunders possessed an "exceptional ability to analyze bird songs by ear," they asserted that relevant "characteristics cannot be accurately determined by ear alone." Even Saunders himself applauded the potential of the sound spectrogram for the study of birdsong. "Writing from the standpoint of what the ear hears," he notes, one's observations "may be more or less different from what the Vibralyzer [the Kay Sona-Graph's forerunner] records" (Saunders 1961, 598). Indeed, like Brand's sound film two decades earlier, sound spectrograms seemed to expose the perceptive limits of human hearing, particularly when it came to the time resolution of rapid birdsongs. As Borror and Reese demonstrated in one of their first spectrographic studies, what had sounded like faint lisps or a single buzzy note to the ear appeared clearly as a series of separate notes in the spec-trographic image. The spectrograph, they concluded in a common turn of phrase, will provide "objective data" that are more detailed and accurate than those obtained by most of the methods heretofore used (Borror and Reese 1953, 276).

It is easy to see the attraction of this new instrument. The spectrograph provided precise acoustic information that could be understood without dedicated training. But what exactly did it allow its ornithologist users to do better than they could by ear? An example taken from the first applica-tions of the sound spectrograph by Thorpe and Marler at the Cambridge University Department of Zoology may demonstrate why it was the spec-trograph that elevated their studies, in Marler's words, "to the level of scientific research." When Marler arrived as a research fellow at Cambridge University, he had already conducted his own large-scale study of chaffinch song, resulting in a classification based on transcriptions made in France, the Azores, the Scottish Highlands, and the English countryside (Marler 1952). Like many field ornithologists, Marler had relied on a self-devised system of transcription to record the chaffinch songs by ear.[5] His study set out to debunk what he thought was a common misconception among field ornithologists, namely that birds display broad regional variety, and thus a song that is regionally characteristic. Instead, he argued, chaffinches maintained a wide range of different song types, song "dialects," that were differently distributed across different regions. He speculated that such variations could be explained by the way individual chaffinches learned their songs (Radick 2007, 247–253).

In a separate study on chaffinch song, the Danish ethologist Holger Poulsen (1951) had recently suggested that it was learned at least in part from adult birds in early adolescence, which would explain convergence between the songs of birds in the same localities. This force of adaptation, and its role in variation, was the subject of Thorpe's spectrographic experiments on the learning abilities of birds (Thorpe 1951). In the seclusion of his laboratory, Poulsen had reared two chaffinches that produced abnormal songs, allowing him to specify some of the song elements that birds developed innately and the stage in their development at which these were modified. In similar fashion, Thorpe (1954) later isolated groups of juvenile birds in aviaries and exposed them in different degrees to the songs of mature singing birds. This enabled him to control different stages of development and establish which parts of their songs the birds already possessed without learning. Crucially, whereas Poulsen had identified all observable changes in song by ear, Thorpe reproduced the songs as spectrograms and compared them visually. As a result, he could follow in detail how the songs developed over time. He found, for instance, that young chaffinches did seem to possess a minimal "blueprint" for their song, which was one of the determinants of its length and form. Other details were clearly learned. Some elements must have been learned in the chaffinch's first weeks, even before it sang itself, supporting Lorenz's concept of "imprint," to which Thorpe was sympathetic. The experiments also confirmed Marler's observations regarding dialects. Since chaffinches refined their song through imitation in their first spring, as they competed for the territory they would occupy for the rest of their life, their songs naturally converged in local, but nonetheless individually distinctive, patterns. However, the sonagrams also demonstrated that even as chaffinch songs matured, they never became completely fixed. Some elements of the bird's song displayed subtle differences between the first and second year. It was these variations, "so minute as to be practically imperceptible to the naked ear," that came to light in spectrographic renderings of the songs' acoustic structures (Thorpe 1954, 468).

Comparisons of spectrographic representations made it clear that individual birds' song was much less fixed than previously assumed. It had long been thought that the distinctiveness of a species' song was a reproductive isolating mechanism. However, spectrographic studies of birdsong increasingly suggested that variation was a much more prominent organizing principle than had yet been acknowledged. Taking stock in 1958, at a symposium on animal sounds and communication at the American Institute of Biological Sciences, Marler (1960) distinguished several biological levels on which variation took place simultaneously. There was, of course, geographic variation, illustrated by the fact that chaffinches of the same species in the Azores sang less elaborately than in Britain. Within a given geography, adjacent populations of birds also seemed to employ various song dialects. Moreover, even within a single population of birds, individual birds consistently varied their song, to such a degree that experienced field observers could distinguish between individuals. Finally, within a single bird's repertoire, there could be hundreds of song themes. Zoologists were now beginning to glimpse how the balance between a birdsong's individuality and its conformity to local, geographic, and species-specific patterns was fine-tuned through adaptation and selection. In the mid-1950s, the Cambridge University ethologists had identified some of the mechanisms by which chaffinch songs varied, but as yet their precise behavioral purposes were unclear. A first step in answering these questions was to compile detailed descriptions of the occurrences of variations, on all of these levels and for more species than the chaffinch alone—and the best instrument for that task seemed to be the sound spectrograph.

At the end of the 1950s and in the early 1960s, comparative spectrographic analyses were widespread. In Cambridge, Thorpe (1961b) and some of his graduate students such as Richard J. Andrew (1957) had begun to inventory and study the vocalizations of buntings. The Madingley field station had by then expanded, and in 1960 it was officially recognized by the university as the "Sub-Department of Animal Behaviour."[6] Marler, who moved to the United States to become a professor at the University of California at Berkeley in 1957, continued with new model animals for studying vocal learning, such as the white-crowned sparrow and the zebra finch. Marler's graduate students, such as Masakazu Konishi and Fernando

Nottebohm, went on to examine aural feedback on song learning and to pioneer the neuroethology of birdsong, while Marler himself embarked on field studies of primate communication. Elsewhere, the comparative approach was taken up by Donald Borror (1959, 1961) in Ohio, Peter Kellogg and Robert Stein (1953; see also Stein 1956) at Cornell University, and Wesley Lanyon of the American Museum of Natural History, together with William R. Fish (1958).[7] Traditional ornithological studies had restricted their focus to the song of a small number of individuals or a small population at most. These papers, in contrast, collectively shifted their attention to the inventory and comparison of song repertoires of several populations at once. The spectrograph also obliged them to organize their investigations differently. Whereas the naturalists discussed in chapter 2 had tended to prefer elaborate and varied songs, these recordists necessarily turned to birds whose repetitive, abundant, and short vocalizations could be represented most effectively in a sonagram. In the 1950s and 1960s, the spectrograph could analyze and represent fragments of only 2 to 4 seconds, and shorter, repetitive songs were better suited for representation. Thorpe's favored experimental subject, the chaffinch, had been selected not only because it bred well in captivity, but also because it produced "a complex but not too elaborate phrasical song of medium frequency range and convenient length." It also displayed local peculiarities, and as such allowed representative sampling of song variation (Thorpe 1954, 466).

Although variation was not discovered using spectrographic analysis, it greatly facilitated and accelerated the making of representation that benefited studies of variation on a large scale. It allowed the analyst to focus not solely on specific features or parameters, such as pitch or duration, as musical recordists had done in the past. Instead, researchers adopted the spectrograph developers' view that "what the sonagrams show best is pattern."[8] Spectrographers would examine and compare the visual print or "structure," as they called it, of a song fragment, with regard to its shape and spacing. This is illustrated, for instance, in a glossary of song components compiled by Marler and his Berkeley student Miwako Tamura, which distinguishes between "notes," "phrases," and "syllables" on the basis of their visual shape alone (Marler and Tamura 1962). Distinguishing a song thus visually, rather than audibly, in its smallest units of analysis allowed spectrograph users to make detailed comparisons of a song structure (figure 5.3). But it also introduced new problems. Terminology, for instance, became almost as controversial an issue as it had been in the early decades of the twentieth century. According to eyewitnesses, a session on bioacoustics terminology at the International Ornithological Congress in

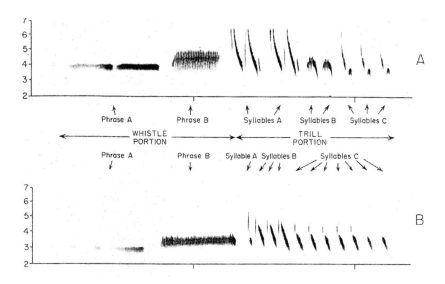

Figure 5.3
Visual model for dividing song structures into analyzable components.
Source: P. Marler and M. Tamura, "Song "Dialects" in Three Populations of White-Crowned Sparrows," *Condor* 64 (5) (1962): 369. Reproduced with kind permission of the American Ornithological Society.

1962 prompted "almost a riot"; after it became clear that "divergent and strong opinions came from every quarter … it was obvious that we are still in a very elementary stage on that score."[9] In the absence of agreement on a standard terminology, individual researchers and groups devised their own terminologies, taking their cue from music or linguistics, which resulted in confusion that would remain unresolved for at least a decade (Baker 2001, 8; Shiovitz 1975).

Producing Sound Spectrograms

Although the sound spectrograph had simplified methods of reading variations in song structures, producing useful spectrograms was not straightforward. It required selections to be made, settings to be decided on, and images to be reproduced and eventually printed. A commercial sound spectrograph such as the Kay Sona-Graph posed a number of trade-offs: accuracy in frequency measurements had to be sacrificed, for instance, for precision in measurements of time. This elicited from Peter Kellogg at Cornell the observation that "the important thing about the [sound spectrograph] is that no one trace shows everything. The technique used in an analysis is very dependent upon the characters you wish to show or emphasize."[10] As a result, conventions for the appropriate production and reproduction of information in a spectrogram differed significantly and controversially.

Some users, for example, found that the amount of information contained in a spectrogram hindered a good understanding of its spectral properties. In a note on the illustrations to his 1961 monograph on birdsong, Thorpe (1961a, xii) explained that "for many purposes of the student of bird behaviour, and for the general ornithologist, sound spectrograms contain a great deal more information than is relevant to the particular point at issue." He had often found it "advantageous to reproduce [the spectrograms] in a somewhat diagrammatic and stereotyped form which draws attention to the main items of information without confusing the picture with a great deal of irrelevant detail." To highlight relevant structures and mark significant patterns in a song phrase, biologists at the Cambridge University Department of Zoology customarily resorted to tracing the original spectrograms with pen and ink. Alternatively, they would sometimes reproduce spectrograms as high-contrast photographic plates. This not only facilitated the reproduction of the spectrograms in print but, as Marler and Tamura (1962, 369) noted, also allowed the researchers to retouch the copies with white paint to mask traces of "background noises" that had been picked up by the microphone (figure 5.4). This treatment of spectrograms was common in bioacoustics publications: from the late 1950s onward, soft-toned photographs were gradually replaced by high-contrast reproductions and ink tracings. In fact, as late as 1979, editorial guidelines of the ornithological journal *The Condor* required contributors to prepare spectrograms on high-contrast film to produce a strictly black-and-white copy, which significantly cut the cost of reproduction and allowed "extraneous sounds to be erased with paint or white correction fluid" (Thompson 1979, 220).

In "sampling" down bird voices to lower definitions and filtering out extraneous noises, bioacousticians' concerns resonate with those of communication engineers such as Franklin Cooper and Alvin Liberman at Haskins Laboratories, who a decade earlier had relied on a similar technique for painting over simplified spectrograms to determine the minimum parameters by which signals could be stored, retrieved, and transmitted without significant losses in information.[11] But where these engineers' drawings had entailed an experimental attempt to explore visible sound patterns, bioacousticians typically had already acquired a strong sense of which traces could and should not be eliminated in order to preserve their

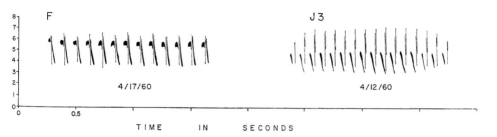

Figure 5.4
Ink tracing by hand of a spectrogram.
Source: P. Marler, M. Kreit, and M. Tamura, "Song Development in Hand-Raised Oregon Juncos," *Auk* 79 (1) (1962): 15. Reproduced with kind permission of the American Ornithological Society.

essential features. In this delicate process, contrast and sharp or faint features were adjusted in part by relying on the author's auditory experience. Rachel Mundy (2009, 210) has likened the technique to calligraphy. This is not only apt with regard to the aesthetics of the trace, as sound spectrograms (particularly those that had conventionally been produced with the wideband setting) looked as if they had been drawn with a wide-bladed pen, but also because of the care and skillful penmanship required of the tracer in this process.

The analogy with calligraphy is fitting in another, unsuspected way, as well. In 1961, Thorpe and his assistant Barbara Lade proposed a method for turning sound spectrographic traces into an objective notation. The sound spectrograph was incomparably more precise and objective than earlier attempts at notation, they pointed out, but it also contained more information than necessary for the student of birdsong. Evidently aware of Kopp and Green's attempt to teach visible speech based on spectrograms, the authors drew on the extensive libraries of the BBC and Cornell University to develop a series of symbols for different song types, based on a diagrammatic rendition of their spectrographic patterns. By halving the timescale and doubling the frequency axis, the patterns resembled long curved brushstrokes of different thickness (see figure 5.5). Such a notation, they argued, would be accurate, yet also capable of being read and used by the field student, thus bridging the gap between laboratory analysis and field experience. By eliminating distracting information to show just "the essence of the pattern," the symbols would allow users to recognize a sound by its spectrographic shape, just as birdwatchers could recognize a distant bird by its shape in flight.

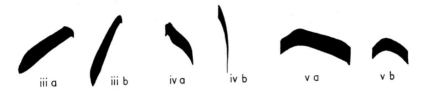

iii a   iii b   iv a   iv b   v a   v b

FIGURE 5.   Simple notes of longer duration: a. sonogram, b. notation. i, Red-eyed Towhee *Pipilo e. erythrophthalmus.* ii, Whitethroated Sparrow *Zonotrichia albicollis.* iii, White-eyed Towhee *P. e. alleni.* iv. Willow Warbler *Phylloscopus trochilus.* v, Icterine Warbler *Hippolais icterina.*

Figure 5.5
Thorpe and Lade's diagrammatic notation based on sonagrams.
Source: W. H. Thorpe and B. I. Lade, "The Songs of Some Families of the Passeriformes. I. Introduction: The Analysis of Bird Songs and Their Expression in Graphic Notation," *Ibis* 103a (2) (1961): 238. Reproduced with kind permission of the British Ornithological Union.

Yet in spite of the appealing simplicity of Thorpe and Lade's idea, an actual spectrographic notation never caught on—at least not with users in the field. Even though the spectrographic traces had been radically simplified in comparison with Kopp and Green's failed experiments in visible speech a decade earlier, the patterns continued to place an insurmountable strain on the reader. This was due, at least in part, to the linear scale with which spectrograms represented measures of frequency, which did not map directly onto the (logarithmic) scale by which a subjective experience of pitch is typically represented. This distortion made it difficult for field observers to translate spectrographic symbols back into their own auditory experience. The idea of reducing a spectrogram's visual complexity to a schematic diagram did prove its worth, however, in laboratory analysis, as a way of categorizing hundreds of song fragments and figuring out how these patterned in sequences.[12]

Even if such diagrammatic renditions of spectrographic sound did not permit an intuitive legibility per se, they had a clear analytic advantage. By manipulating the spectrogram, analysts achieved far greater control over the audible phenomena themselves, allowing them to distinguish between irrelevant "noise" and the "sounds" they deemed scientifically interesting. The spectrogram enabled such distinctions to be made more precisely than the parabolic microphone, more selectively than a crude band-pass filter (that cut off any frequencies below a certain range), and more economically than the soundproof rooms that Thorpe and field station curator Robert Hinde had introduced in previous years (Thorpe and Hinde 1956). These rooms had been installed primarily to isolate the birds themselves from the acoustic ambience of the Madingley Ornithological Field Station. To investigate what part of their song was instinctive and what part was learned, it was necessary to control the kinds of sounds to which they were exposed and exclude any vocalizations of outside or neighboring birds. However, by artificially filtering out such environmental noise levels, the soundproof rooms also contributed to an ideal of the spectrogram as free of exposure to sounds that analysts would discard as extraneous. In spectrography, "noise" came to refer interchangeably to acoustic interference and scrambled information, both of which rendered the communication of essential spectrographic features unintelligible. Just as the parabolic close-up recording aimed to reduce unnecessary background sounds, these ink tracings and high-contrast photographs placed a spectrographic sound against a white and therefore equally "mute" background (Bruyninckx 2012). As such, spectrographic tracings served as a powerful filter to distinguish patterns from noise.

Yet the attempts by Thorpe and his former students to enhance the original spectrogram's legibility for analytic purposes were not embraced by all. Some reviewers of Thorpe's 1961 monograph simply requested more technical detail on his methods of producing sound spectrograms (Stonehouse 1963); others, particularly in the United States, received the book and especially his casual note on the reproduction of his spectrograms with surprise. Ohio bioacoustician Donald Borror and Cornell ornithologist Robert Stein prepared critical reviews.[13] Peter Kellogg, likewise, "didn't think much of [Thorpe's] reasons for tracing spectrograms rather than reproducing them as they are. It seems that such a method as he uses must always include any opinions the tracer must have about the important or unimpotance [sic] of some character."[14] He explained:

> [Thorpe's] idea that, for economy and perhaps for neatness, it is a good idea to trace spectrograms rather than to reproduce them directly, is a technique which I most seriously question. In tracing a spectrogram, so as to clean it up, and also so that it may be reproduced as a line drawing, it is almost impossible to keep from changing the picture a little so as to make it more precisely fit your ideas. This results in emphases which were not present in the original and in the omission of everything which you consider to be, but which may not be, an artifact. In one instance, this technique had led to the inclusion of material not in the original.[15]

Kellogg instead advocated using as little intervention as possible: "I am in favor of publishing spectrograms as they come from the machine rather than to trace them so as to emphasize the pattern and eliminate any details which the author considers at the moment to be of no importance."[16] He also admonished his colleagues and research associates to take reproduction seriously: "I would like to see the spectrograms ... *authentically* reproduced so that they could be used for study with the same confidence as one could use the originals. It might be necessary to reduce some of them slightly, but this should not hurt them much."[17] The extent of his concerns is illustrated by deliberations with colleagues, and the printers of publications to which they contributed, on how best to achieve such authentic reproductions. Results even differed significantly with different printing techniques, he found, and printers were often "amazed to find that I preferred [the spectrogram] to be grey."[18] In this context, well-reproduced spectrograms were regarded as a marker of professionalism. One of the Cornell research associates, L. Irby Davis, himself a retired civil engineer who had become skilled at using the sound spectrograph, complained, for instance, about a series of spectrograms accompanying an article in *Scientific American* because they would reflect poorly on the lab's reputation: "Everything else, the drawings, paintings, and the photography are of excellent professional quality. But the spectrograms are terrible ... . Since the Lab is known all over the world as the leading place for sound work on birds it will be a bit hard to explain how such bad work was permitted to come out from the institution."[19]

At the root of this uneasiness with authentic reproductions was a concern that biologists lacked sufficient expertise in acoustics and electrical engineering. This concern was particularly strong among Cornell bioacousticians: like Kellogg, they often complemented an interest in zoology with a background in physics or electrical engineering. They worried that commercial sound spectrographs had been elevated to a kind of out-of-the-box solution for the study of animal vocalizations, and that their serious mechanical and electronic difficulties might go unnoticed by the inattentive, untrained user. Summarizing this concern, Crawford H. Greenewalt (1968, 8), another Cornell associate, pointed out that although the Sona-Graph had been a perfect tool for the analysis of human speech, birdsong spectra were beset with ambiguities that could be more or less serious depending on the purposes of the experiment and the acoustic characteristics of a song. Those ambiguities were spelled out most concretely by Davis, who insisted that without an adequate knowledge of acoustics, biologists would be prone to mistaking artifacts for actual phenomena (or vice versa). In a series of technical papers, he proceeded to show how frequency-modulation and amplitude changes introduced spurious harmonics, time-frequency smear, or displaced scales into the analysis (Davis 1964).

His uncompromising scrutiny of published spectrographic analyses led to several critical exchanges not just within the Cornell department—among others with converted behaviorist Robert Stein—but also with Cambridge University ethologists like Thorpe and Marler, whose work had meanwhile set the standard in the field.[20] Rather than let such unease simmer, however, Kellogg sought to resolve the matter in characteristic manner; he enlisted Marler, along with collaborators such as William Gunn and William R. Fish on the editorial board of a new quarterly that he launched at Cornell, *Bioacoustics Bulletin*. Distributed among about three hundred recipients worldwide, the bulletin focused on technical discussions of methodology and instrumentation (including a series of papers on the possible pitfalls of spectrographic analysis), thus bringing a young community together under the banner of bioacoustics and deliberating on new standards for its analytic practice.

Subdued though they may have been, such skirmishes illustrate the variety of practices that had emerged around the sound spectrograph by 1960, and the different gestures of objectivity that their users had developed in the preceding decade.[21] For critics of Thorpe's or Marler's adjustments, their spectrograms seemed to permit a dangerous interpreters' bias in the analysis. Marler and Thorpe, for their part, trusted the instrument's ability to produce an objective image of sound, but found that the cluttered image and incidental detail compromised effective analysis. They thus relied as much on an informed and experienced analyst as they did on technological means, to distill from the noise a more sophisticated and distinct image. This discourse recalls what Daston and Galison term *trained judgment*, which emerged as a mid-twentieth-century supplement to the doctrine of self-elimination in mechanical objectivity and left ample room for skilled interpretation and expertise. The ideal of trained judgment did not reject objective instruments outright but, unlike mechanical objectivity, did envisage a role for informed evaluation and cultivated perception, alongside the protocol-based image, to distinguish salient and significant structures, categories, or patterns. The Cambridge University zoologists had developed a spectrographic practice that permitted, or even required, a trained observer to order complex acoustic information for the reader. By relying only on the indiscriminate procedures of the sound spectrograph, they reasoned, the biologist would risk obscuring the very structures and patterns of variation that had become of central analytic interest to bioacousticians and ethologists. As the next section will show, this regime extended not only to looking for spectrographic traces, but also to listening for aural patterns.

Aural Patterns

The sound spectrograph changed the ways sound recordings were analyzed. Yet in the field, listening continued to play an important role. In
1966, a new field guide, *Birds of North America: A Guide to Field Identification*,
included sound spectrograms for depicting songs and calls, but like Thorpe
and Lade's proposal for a concise spectrographic notation, the idea failed to
catch on initially. One otherwise highly positive review judged the practical use of sonagrams as an aid to field identification very limited. Not only
was considerable practice needed to be able to read the sound spectrograms,
but when such proficiency was acquired, he noted, "I doubt if anyone
can really imagine or 'hear' a new song simply by reading its sonagram"
better than through a time-honored method of verbal description (Keith
1967, 253).

That doubt was shared widely. Even scientific papers in *Animal Behavior*
or *The Auk* complemented extensive spectrographic with verbal descriptions or syllabic notations of a bird's vocal behavior to give an impression
of how the sound might seem "to the human ear." This phrase in particular
flagged a deliberately subjective and descriptive account, evoking information that could not be conveyed otherwise. The persistent presence of the
observer's ear is illustrated by British ornithologist Derek Goodwin's comparative analyses of bird vocalization behavior. In a series of descriptions of
blue waxbill calls, he noted that its contact call, "a loud, clear, high-pitched
'tseep-tseep' or 'sweet-sweet' with a somewhat interrogative tone," did not,
"to my ears, usually sound squeaky." And although this call might easily
be confused with that of a related species, "the experienced ear can usually
identify the caller." For other calls, "I cannot distinguish by ear which of
the three forms is calling" (Goodwin 1965, 287–290). Although recordings
had often been dutifully analyzed using a sound spectrograph, authors typically included their aural experiences in the field to provide a perceptual
marker for other ornithologists.

Skilled listening also mattered in the field when conducting playback
experiments or collecting observations of behavior in the field. Observers
listened not only for whether a bird responded or not, but also for what
exactly the response was. In an investigation of cardinals' adjustment and
coordination of their song patterns to the song of neighboring birds, carried out by playing back prerecorded songs, McGill University biologist
Robert Lemon noted that "the data was recorded by hand after identification of the songs *by ear*. This method is feasible with cardinals because of
the relative simplicity and stereotypy of their patterns of song" (1968, 158;

emphasis added). In a later study, exploring the statistics of variations in sound patterns, Lemon and his coauthor noted that although all the songs had been dutifully analyzed with a sound spectrograph, much information, "especially relating to the sequences of different song types, was gathered by listening to the birds sing and then recording the data in a notebook" (Lemon and Chatfield 1971, 1).

But aural experiences were not limited to field observations alone; they also played a role in the organization and interpretation of sound spectrographic data itself. Analysts occasionally reported using a kind of discriminative listening when considering song types in the laboratory. Whereas most papers did not give details on how their data had been classified, some explicitly invoked aural experience alongside spectrograms as an aid to interpreting and comparing sound fragments or classifying material. When classifying Carolina wren song phrases for the variations in number, length, and notes that they displayed, Borror had evidently relied on the sound spectrograph. But when drawing up his classification of song phrases, he encountered difficulties in defining the beginning and end of a song phrase on an image. This was important, because "a different delimitation of the phrases would for most songs result in a different classification" (Borror 1956, 223). Borror addressed the problem by combining spectrographic imagery with listening to the recordings played at reduced tape speed. In a comparable vein, at Cambridge University, Thorpe noted that recorded sounds could be studied by a variety of means: even if the sound spectrograph seemed "by far the most valuable method," "play-back at the lower speeds is an enormous aid to the ear, particularly with sound patterns ... having extremely rapid repetitions and relatively high frequencies. ... It sometimes happens that comparison of songs of related species at decreased speeds brings to light resemblances which would otherwise have escaped notice" (Thorpe 1958, 542). In such cases, the sound spectrograph could again help to verify observations made by ear.

In other laboratories, too, discriminative listening continued to play an important (albeit a rather inexplicit and context-dependent) role, even when sound spectrography had become firmly established as the standard instrument in birdsong biology. British biologists Peter Slater and S. A. Ince (1979), for instance, reported that in creating a classification of the songs they had collected in the field, they had relied on their own experienced listening. Preparing spectrograms of each song type, they had found that "with practice many of the more distinctive song types could be identified by ear" (Slater and Ince 1979, 148). Identifications were made even easier by listening to slowed-down recordings with the typical sonagrams at hand. Checking their identifications by spectrographic analysis, the authors reported finding a host of reliable features that could help distinguish between some similar-sounding song types, and failing to find them for a few others. The balance between listening and looking, therefore, depended on the task at hand and on the types of sound being listened to. In some cases, Slater and Ince continued, the differences between song types appeared slight on the sonagram, but the types were nevertheless classified as separate on the basis that certain differences in quality had been "immediately recognizable in the field on the first occasion that song type Y was heard" (Slater and Ince 1979, 157). For another group of song types, "there is no doubt that they should be regarded as distinct because the differences in form between them are consistent," even though the authors had been "unable to separate them reliably by ear" (Slater and Ince 1979, 157). Trained judgment thus operated at two levels. First, it helped the recordist to recognize similarity relations, family resemblances, and distinctive types in the diversity of records; in a second and related way, it involved a keen awareness of the exact moments when particular judgments could dependably be relied on. A trained observer was able to tell, for example, when aural impressions could authoritatively trump the visual evidence suggested by a spectrographic print.

Slater and Ince also found that some of their observations on chaffinch song variation did not match Marler's (1952) earliest study on the subject, made as a student, which he had completed entirely using naturalist standards—by ear and pencil. The discrepancy in some observations, the authors noted, was probably due to the fact that Marler's original collecting and analysis had been carried out by ear alone: whereas most end phrases of a chaffinch song might reliably be captured by ear, its extremely rapid trill usually displayed differences that were particularly hard for the human ear to notice without a sound spectrogram. Thorpe and his colleagues (1972, 134), in turn, acknowledged that "any study of sound presupposes the use of the ear, … if the task is not to become cumbersome and time-consuming out of all proportion to the results achieved." The task of spectrographing and comparing hundreds of records could sometimes be performed more efficiently by relying simply on the ear. But this dependence on aural experience must always be qualified by a visual/mechanical record, since "human aural perception … tends to reduce disorder to a preconceived order and may categorise within the familiar apperception masses those aspects of a study which, when considered with complete objectivity, may be most likely to lead to new notions and syntheses" (Thorpe et al. 1972, 134–135). For that reason, they suggested, a study of sound patterns should generally "begin with aural classification and continue with suitable methods of mechanical analysis" (Thorpe et al. 1972, 135).

In this sense, the trained listener presupposed by these spectrographic studies was different from Witmer Stone's trained musical listener (chapter 2) or Albert Brand's all-too-human subjective listener (chapter 3). Listening did not feature here as a distinctive skill, nor did it stand in direct opposition to the spectrographic image, but rather it served to remedy the inefficiencies and deficits inherent in the spectrographic process. This approach was usually not made explicit, but rested instead on a perception cultivated through experience with the sounds under study. While the experienced listeners implied in these papers embraced the mechanical objectivity of a spectrographic image, they also developed an intuitive understanding of how to make objective records work most efficiently. Despite the assumption of congruity between aural and visual patterns that underlay the sound spectrograph's original design, some aspects of birdsong could not be captured by the spectrographic image alone, requiring the analyst to rely on the ear to adjust, categorize, and classify sound—always in conjunction with the mechanical record.

Musical Spectrograms

However persistent listening's role in the routine of spectrographic analysis, the authority of the trained ear itself was carefully delineated. The restrictions on the domain in which trained listening could be deployed were most clearly articulated in the 1960s and 1970s, when a small number of researchers expressed discontent with the standard spectrogram and began to codify bird sound in new, unconventional ways that, surprisingly, used music as a reference.

Among these were several biologists and musicians who orbited around Thorpe at the Cambridge University Sub-Department of Animal Behaviour. Thorpe himself had begun to develop an interest in musical notation as an analytic tool. By 1962, his attention had shifted from chaffinch song variations to the ritual "duet" vocalizations of male and female shrikes, and he found that to appreciate exactly how these birds developed their song patterns in interaction with each other, it was most effective to note down their variations musically. When sounds were used for communication, and particularly when they were of a musical nature, he argued, one must consider the ear rather than a mechanical instrument as the analyzer. It was, after all, fair to assume that "since its essential structure is similar, the avian ear is subject to the same distortion" (Thorpe et al. 1972, 135).

To that end, Thorpe surrounded himself with musically expert inter-locutors and collaborators. At Cambridge University, musicologist Thurston Dart taught him techniques for the musicological transcription of folk music. Music graduate and professional musician Joan Hall-Craggs joined the sub-department with a project on blackbird song. Thorpe encouraged composer Trevor Hold to develop a musical notation that was better suited to the representation of birdsong. He also maintained regular correspondence with Myles North, the Cornell lab research associate stationed in Kenya. In 1950, North had published a short technical paper advocating for the notation's continued importance and proposing that ornithologists use a stripped-down musical notation for identification and field observation (an example is shown in figure 5.6). Acknowledging the recent advances in "accurate" birdsong recording and their "physical analysis," he argued that musical transcription as a parallel and allied line of study would be mutually profitable (1950, 101). North helped Thorpe refine his musical transcriptions—the results of which were published in *Nature* in 1965; in a second paper in 1966, Thorpe used spectrograms instead.

These notations would not find wide acceptance in the field of birdsong biology. Yet they and their reception do offer important insights into the definition of trained listening. Thorpe and his Cambridge colleagues adopted musical transcription techniques because they were

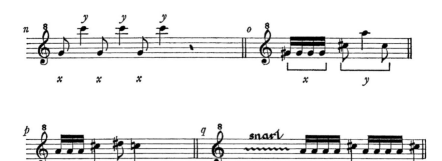

Figure 5.6
Musical notation of duetting shrikes by Thorpe.
Source: W. H. Thorpe, "Ritualization in Ontogeny. II. Ritualization in the Individual Development of Bird Song," *Philosophical Transactions of the Royal Society of London B: Biological Sciences* 251 (772) (1966): 355. Reproduced with kind permission of the Royal Society of London.

interested in a very specific research problem. They did not want to show *that* birds developed their song, but to analyze exactly *how* and according to what principles they did so. To demonstrate such developments qualitatively, the researchers found, musical notation did the job just as well as the shapes of acoustic structures produced by the sound spectrograph. Thorpe wrote that to study how birds elaborated their song patterns through ritualized interactions, it was "essential to put [the songs] first into staff notation so that I can get certain things clear for myself and have the songs in a form which I can show to musical people and get their help and advice."[22] But by transcribing their analyses in musical notation, these researchers did invite in again a particularly thorny issue that had been left far from resolved in the late 1920s. With the general adoption of the sound spectrograph, biologists of birdsong had moved away from the elaborate and—to the human ear—often aesthetically pleasing song patterns that had led early-century naturalists to consider their musical nature and intention. In search of variability, they had focused on particularly brief calls and songs. Within the frame of ethology, moreover, attention had shifted to consider their behavioral functions and the kinds of information they communicated. This was particularly evident in the surge of field studies that tested a species' response to prerecorded vocalizations that were played back at them or in a concern with the syntax of their signals (Falls 1992). In some cases even, this shared interest in communication as well as a shared investment in spectrographic analyses had brought ethologists and linguists to consider to what extent such vocalizations could be considered a language.[23] If biological acousticians in the 1960s looked to relate bird vocalizations to a reference point in human behavior at all, therefore, it was to language much more than music.[24]

Hence, when Thorpe and his collaborators introduced the songs they had recorded in musical notation, they were careful to point out that these did not imply necessarily that birdsong itself was inherently musical, although they did bring to light several aesthetically pleasing patterns, suggesting "musical inventions" that seemed to transcend biological necessity. Conceding that "whether the musical tonal system employed and the manner of using it provides any justification for assuming the beginnings of a true artistic ability is still an open question," Thorpe (1966, 357) argued that it would be dishonest to suggest that the biological theories at present available offer a complete explanation for all bird vocalizations. Similarly, when Hall-Craggs recorded and transcribed blackbird song in musical score, she was seeking to conceptualize how the bird reorganized its song patterns in ways apparently beyond a purely biological function. As a musician "with the sole qualification in the study of birdsong of being trained to listen to detail," she added, "it would be presumptuous to try to draw conclusions from this analysis" before she continued to point out that the biological functions of blackbird song could conceivably be mixed with an aesthetic sense, even if this was not an easy hypothesis to accept as yet (Hall-Craggs 1962, 294). In an attempt to dodge the unavoidable accusation that they might be anthropomorphizing birdsong, Thorpe and Hall-Craggs stressed that their methods had not surrendered accuracy or objectivity. Even for musical transcription, they insisted, their trained ears had depended on mechanically objective records. Thorpe and Hall-Craggs had combined their interpretation of musical transcriptions with spectrograms of the records, and aural transcriptions had only been made on the basis of tape recordings, allowing them to slow down the original sounds to half or even one-sixteenth of the original speed and transcribe the rapidly uttered bird-songs by ear in minute detail, with a corrective factor for transposing them back to their original speed.

The technique was not unique to Thorpe's laboratory. By the 1960s, there were several semiscientific popular gramophone publications available on the market that employed exactly this technique to uncover the "hidden" nature of birdsong, with titles such as *Music and Bird Songs: Sounds from Nature, With Commentary and Analysis* (put together by Cornell University's Peter Kellogg and CBS Radio broadcaster James Fassett [1953]), *The Birds World of Song* (published by the ornithologist couple Hudson and Sandra Ansley [1961] in the Folkways Science series), or the later *The Unknown Music of Birds* (by the obscure Hungarian musicologist Peter Szöke [1987]). Their authors used the medium of the gramophone to reduce a selection of bird vocalizations to fractions of up to 1/64th of their original speed. In doing so, they aimed to transpose them to a "human scale," or rather, present the vocalizations as they were presumably known to the birds themselves. These were records intended for a general audience—Fassett's record had originated as an experimental artistic composition broadcast by CBS during the intermission of the New York Philharmonic—but they all couched their liner notes in some scientific terminology. While the cover of *Music and Bird Songs* featured a wave analysis produced by bioacoustician William Fish, the Ansleys and Szöke both presented their approach as the aural equivalent of the microscope—a way to zoom in on sound fragments and overcome the "physical limitations" of the unaided ear—thus mobilizing the same trope that Marler later proposed for the sound spectrograph. At the same time, these authors also resumed an analytic practice that many proponents of the sound spectrograph had expressly sought to avoid. In fact, the Ansleys' analytical frame of reference was closer to the artisanal transcription techniques of Ferdinand Mathews or Aretas Saunders than to spectrographic visualizations, which they used to warn against strict distinctions between human and animal music. Likewise, Szöke, who is now considered a precursor to investigators in zoomusicology, fought a rearguard action against the spectrograph, which he proclaimed inadequate to study matters of intonation and expression in birdsong. The point is that while the authors of these records embraced listening techniques that were ostensibly very similar to those employed in the laboratory of the Cambridge University zoology department, they also differed in an important way. Thorpe and his collaborators in the mid-1960s and 1970s listened to their recordings to fine-tune, not to replace, spectrographic analysis. Above all, they hoped to reconcile both of these positions.

The Cambridge University Department of Zoology was not alone in seeking such reconciliation by pairing spectrographic analysis with musical conventions. One ornithologist, Joe Marshall Jr., converted the linear frequency scale conventionally applied in the spectrogram into a logarithmic frequency scale of musical octaves (figure 5.7). Spectrographers usually preferred the linear frequency scale because it makes use of an easily measurable unit and stretches the frequency spectrum equally, thus showing as much detail in the higher intervals as it does in the lower. A logarithmic frequency scale, in contrast, squashed all that detail but, so its proponents

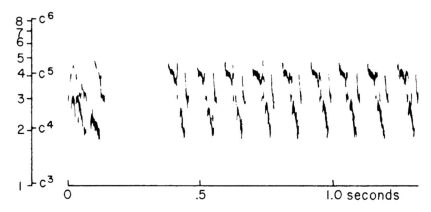

Figure 5.7
Adapted spectrogram fitted with logarithmic scale and musical octaves.
Source: J. T. Marshall Jr., "Voice in Communication and Relationships among Brown Towhees," *Condor* 66 (5) (1964): 346. Reproduced with kind permission of the American Ornithological Society.

claimed, allowed readers to judge better how it sounded. This was an important difference; after all, Marshall (1964, 347) explained, "it is pitch and not
frequency to which our hearing and that of the birds respond in nature."[25]
In his view, bioacousticians' reliance on frequency as a unit of analysis had
reified a focus on spectrographic variations that were often barely perceptible to the ear of the analysts *or* of the birds, "even granting that birds'
ears are much more sensitive than man's" (Marshall 1964, 354). The conventional scale, in other words, "resembles nothing in the real world,"
whereas the logarithmic scale "like a musical score, constitutes a universal
'language' or symbolism by which sounds can be recognized visually by
their shapes on a graph" (Marshall 1977, 150). At Cambridge University,
Joan Hall-Craggs likewise lamented birdsong biologists' strong attachment
to a spectrographic standard that threatened to dissociate sound patterns
from the natural world. Frustrated to observe that auditory stimuli were
often discussed exclusively in the visual terms of the sound spectrogram,
she suggested developing sound spectrograms that, "while maintaining the
objectivity of the analytical process, are (1) presented in a form more accessible to the auditory imagery of readers and (2) comprehensible in musical
terms" (Hall-Craggs 1979, 186). Like Marshall, Hall-Craggs tried to enhance
the mechanically objective rendering of a sound by superimposing musical
conventions onto the sonagram (figures 5.8 and 5.9). This orthographic
innovation was especially valuable for fieldworkers, she found, as it allowed
the reader not only to examine a sound pattern in detail but also to understand it acoustically and thus to rehearse and memorize it for later recall in
the field.

In ornithological circles, Hall-Craggs's proposal for a spectrogram that
would also be accessible to fieldworkers found a welcome reception. In the
preceding years, she had been asked to assist E. Max Nicholson in editing the voice sections of the renewed handbook *The Birds of the Western
Palearctic*, a mammoth nine-volume tome whose first volume appeared in
1977 (Cramp and Simmons 1977). Taking lessons from earlier field guides,
they introduced a new type of sound display to complement the familiar
sonogram to represent species vocalizations, namely the melogram. The
melograph, the device that produced such melograms, had been developed in the early 1960s by Swedish researchers as a way to measure cardiac
and respiratory activity, and refined by musicologists in search of a way
of objectively analyzing instrument performances (Hjorth 1970, 5). It had
been designed to provide precise information on only the fundamental frequency, not to display its entire acoustic structure (such as the overtones,

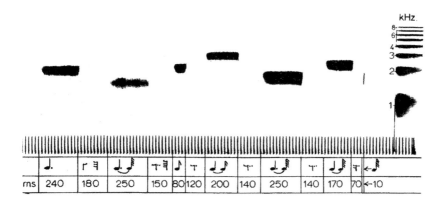

Figure 5.8
Sound spectrographic analysis of a wren song, with logarithmic scale and musical score superimposed.
Source: J. Hall-Craggs, "Sound Spectrographic Analysis: Suggestions for Facilitating Auditory Imagery," *Condor* 81 (2) (1979): 190. Reproduced with kind permission of the American Ornithological Society.

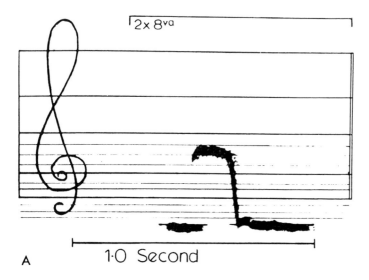

Figure 5.9
Musical staff made by a sound spectrogram, superimposed with microtone intervals.
Source: J. Hall-Craggs, "Sound Spectrographic Analysis: Suggestions for Facilitating Auditory Imagery," *Condor* 81 (2) (1979): 189. Reproduced with kind permission of the American Ornithological Society.

as rendered by the spectrogram), and to present it on a logarithmic scale as in musical notation (Hjorth 1970, 26).

But whereas reviewers welcomed the displays as a solution to the thorny problem of song description in field guides, biologists of birdsong were more critical of her proposal.[26] In a brief commentary in *The Condor*, Edward Miller (1980, 234) objected that although Hall-Craggs's suggestions might have "heuristic value," they were "clearly biased." "A musical (or other) notation of birdsong implies a particular kind of structure or order," he warned, adding that "we must be careful not to assume that such order exists, just because of the system of notation used." If Hall-Craggs had identified musical qualities in the sounds she studied, he concluded, she had focused on an insignificantly small fraction of all possible animal sounds. The commentary echoed a critical analysis of Thorpe's recent "musicological" work by two other biologists, Charles Dobson and Robert Lemon, who checked their own data on frequency intervals in white-crowned sparrow songs alongside Thorpe's published data. They found no significant indication that the intervals Thorpe identified correlated with fixed musical intervals. Thorpe's data were only correct, they argued, because he had used two different scales simultaneously: "by doing so, closer conformity to musical scales could not help but occur" (Dobson and Lemon 1977, 889). Using methods like adapted musical notation, "musicians such as Messiaen have been able to simulate natural bird song with some success. But the use of the standard musical notation commonly employed may lead one to overestimate the musical nature of bird song" (Dobson and Lemon 1977, 890). The problem with musical notation, in other words, was that it implied a preconceived structure that led to subjective projections.

Although this exchange of commentaries did not make significant waves in the field, it throws into relief birdsong biologists' epistemic investments in both the trained ear and the sound spectrograph. By combining spectrographic visualization with musical listening, a select group of Cambridge researchers not only welded together two distinct technologies that mediated and structured the acoustic world in very different ways (linearly and logarithmically) and that had constituted strongly opposed traditions since the 1920s and 1930s (see chapters 2 and 3). They also sought to recover what had initially been the defining feature of the sound spectrograph: a naturally existing language of sound that enabled sound to be represented objectively and transparently, immediately, and with intuitive intelligibility—a visual language, in other words, that brought together the objective physical properties of sound in the laboratory and their phenomenological expression in the field. The concept of visible speech itself proved

difficult to achieve in practice: visual spectrographic patterns were not necessarily congruent with aural patterns, nor were they always immediately
intelligible to the reading eye. However, the notational innovations that
circulated at the Cambridge Sub-Department of Animal Behaviour—Thorpe
and Lade's diagrammatic notation or Hall-Craggs's musical spectrogram—
extend this aspiration. They also connect the Cambridge University biologist in the 1970s to the naturalist in the 1910s. In both periods, students
of birdsong sought to combine accurate, detailed analysis with an intelligibility that allowed sounds to be sensed aurally. In both periods, too,
musical notation divided proponents and critics over its perceived abilities
to communicate sound effectively but within a preconceived and possibly
suggestive format.

Mastery over musical notation had once granted the naturalist the
authority of an expert listener. For its proponents, music still provided a
frame for analyzing, discussing, and experiencing sound. For its critics,
however, musical notation introduced a dangerous bias, one that could
lead readers to misinterpret the score as suggesting that birds' singing
behavior was of a musical nature, or even a musical purpose. With that concern, birdsong biologists employed a frame that had existed at least since
the 1920s, when musical notation became tainted by associations with
anthropomorphism, amateur naturalism, and the arts, more than with the
practice of professional biologists. As Eileen Crist (1999) pointedly demonstrates, in the course of the twentieth century, professional biologists had
been cultivating a highly mechanical idiom that allowed them to discuss
in a technical and detached way those phenomena they had pronounced
unverifiable by the human observer, such as animal mind, emotion, or aesthetic intent. To establish the study of animal behavior as a rigorous science
and ward off any suspicions of anthropomorphism, ethologists thus sought
to purge their observations of any such characterizations of animal subjectivity. Indeed, in bird biology too, by the 1960s, the "validity of drawing
upon human experience in the interpretation of bird song" and the conviction that "bird song has a much *deeper* significance" had been relegated
largely to the naturalist domain (Murie 1962, 181–182). Where birdsong
biologists did consider aesthetic sensibility among birds, they categorized
their specific behavior by a limited subset of species only. By repurposing
musical notation, if only to bridge physical analysis and phenomenal experience, birdsong biologists feared that their proponents invited the human
measure back in as a frame for ethological explanation. If some birdsong
biologists had dismissed notation as subjective, it was not only because it
invited interpretation by a skilled listener, but especially because its means

of transcription opened the door to an animal subjectivity. In contrast, the spectrogram allowed sound to be considered in terms of physical properties only, dissociating sound not only from its phenomenal (and possibly flawed) human interpretation but also from unwanted analogies to such human experience.

Visual Inscriptions

Although the sound spectrograph failed to live up to its promise to connect the "objective" and physical properties of sound with their phenomenology, biologists ultimately embraced it for wholly other reasons. Chiefly, the sound spectrograph was favored for its ability to package sound into traveling inscriptions—what Bruno Latour (1986) famously termed "immutable mobiles." In the agonistic network perspective advocated by Latour and Woolgar (1986), such paperwork is crucial in order to muster together and send off convincing proof to convince as many allies and critics as possible in the absence of the original phenomenon or thing. In their most basic form, these inscriptions enabled unique sound events in the field to be cast in many identical copies and circulated widely, but as Latour points out, inscriptions are not only a matter of duplication. They have a host of advantages that help us understand the importance ascribed to the sound spectrograph.

In the first place, inscriptions are highly *mobile*; because they can be duplicated and printed, they can be circulated more easily around a network of peers. This enabled interpretations of birdsong (whether accurate or not, conflicting or not) from very different places to be collected in a single place—the laboratory or a printed article—without having to return to the specific sound event or its recording. Inscriptions are also *immutable*. Recorded sounds were vulnerable to destruction (erasure) or alterations. Even for mechanical recordings, their reproduction at different locations remained variable, depending on the type, settings, and calibration of the equipment. Paper inscriptions crystallized such variables into a permanent representation, although preserving that information could be cumbersome too. Some types of oscillograms were light sensitive and in order to provide them with the permanence of a black-and-white photographic print, they had to be developed under orange safelight, fixed, washed, and dried (Greenewalt 1968, 13). Spectrograms were less labor- and time-intensive to produce, even though they still required additional work to retouch and reproduce.

Inscriptions are *flat* and hence easier to dominate. Sonic inscriptions, such as the musical staff, the graphic drawing, or the sonagram, have always been rendered as two-dimensional forms. Their use is analogous to that of perspective as a way of transposing an immense three-dimensional building onto paper. These forms projected the dimensions of sound (time/ rhythm, frequency/pitch, amplitude/loudness) onto the flat surface of paper, where they were more easily overseen, cut up, scaled, recombined, or superimposed—in short, controlled. This is particularly evident in the routine practice of tracing images with pen and ink. In the flat dimensions of the image, sounds could be added or removed with an efficiency and economy that, despite innovations such as the parabolic microphone, simply did not exist in the field. Manipulations on paper enabled the researcher to intervene virtually (but none the less effectively) in its soundscape. The cost and effort that went into reducing acoustic interference in the field were much greater than the ease with which researchers could organize and manipulate the soundscapes of that same field spectrographically.

By transposing sounds onto the paper in a standard layout, moreover, very different inscriptions could be made *optically consistent*. This enabled users to represent sounds synoptically, by juxtaposing them in the same plane. In this sense, flat and printed inscriptions work differently from sonic experience, as is illustrated by a sound spectrographic comparative analysis of western meadowlark songs by Lanyon and Fish (1958). The article presents a plate of four spectrograms of a call note (figure 5.10). Although the calls were recorded in different field locations, from Chihuahua to Wisconsin, they could be presented together in one plane and in the same scale. This commensurability of sounds sampled in very different environments enabled these authors to argue that all the notes were identical, regardless of their geographic location. The quadruple comparison of evidence was itself Lanyon and Fish's central argument, made possible only by combining flat inscriptions and adjusting their scale of measurement. This optical consistency allowed researchers to accumulate elements from the soundscapes of dispersed geographies. To understand the additional value of such a representation in the context of bioacoustic research into song variation, one need only imagine the same fourfold presentation of the original recordings, but played out loud uninterruptedly, simultaneously, or even in rapid succession. Although trained listeners could, and indeed often did, try to compare a single pair of samples, none of the published research papers reported analytic benefits from simultaneous listening to four samples, let alone a hundred. Unlike auditory recordings, within

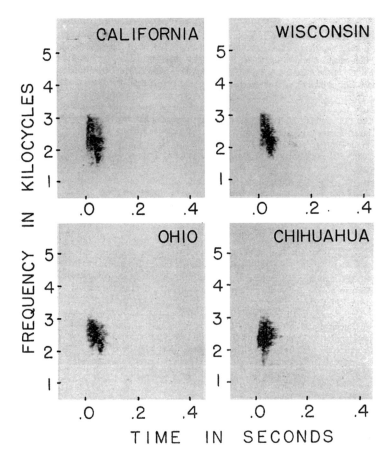

Figure 5.10
A synoptic display of the western meadowlark by Wesley Lanyon and William Fish.
Source: W. Lanyon and W. Fish, "Geographical Variation in the Vocalization of the
Western Meadowlark," *Condor* 60 (5) (1958): 340. Reproduced with kind permission
of the American Ornithological Society.

their flat dimensions the printed inscriptions could be made to "speak"
more easily only by being looked at—one at a time or all at once.

But the meadowlark inscriptions not only allowed researchers to present
dispersed sound segments synoptically. They also enabled the authors to
deal with the sequentiality of sound. After all, these unique sound events
were not only tied to a specific time and place (being ephemeral) but also
happened "in time" (being sequential); a record could preserve the sound
itself, but when it was played back every note was still evidently replaced

by the next, then the next. Whereas listening necessarily takes place *in time* and thus *takes time*, inscriptions stabilize time—allowing the reader to move back and forth or cut across sections more or less at will. It is this stabilization and reversibility of time that matters, more so than the visual organization of sound per se (as may be clear when an analyst is imagined to compare four moving images simultaneously while observing them through a narrow window).

It is, therefore, not the ability to look at sound that made the sound spectrograph and other inscription devices preferable to its study, but what it allowed the researcher to do: namely, to take control over the time and space in which a sound event took place. Slowed-down gramophone recordings and combined sound spectrograms both enabled this kind of control, but evidently to different extents. The inscriptive visualization of phenomena is not an end in itself; it is not imaging itself that guarantees the authority of inscriptions among peers. The bioacoustician realized that a multitude of inscriptions could swamp an observer almost as much as the original sounds would have. For that reason, some first classified the sounds according to their aural impressions, and only then spectrographed typical instances. But the advantage of imaging is that visual representations can be made to cascade with increasing efficiency into ever simpler, more abstracted, and more crystallized inscriptions. As such, a sound spectrogram of birdsong constituted a further step in the processes of sterilization, compartmentalization, and abstraction of sound that began with the focused, directional recordings produced in the field. Spectrograms could now be combined with benchmarks, geometry, and scales and thus more easily be conceived of as a set of mathematical relations. Indeed, confronted with a multitude of recorded sound data, researchers were not content merely to compare spectrographic contours. Lemon and Herzog's (1969) inquiry into the organization of cardinal song, for instance, yielded not just a collection of sonagrams (figure 5.11), but also, and especially, numerous statistical interventions that compressed dozens of individual recordings into a single table (figure 5.12).

The sound spectrograph served this cascade particularly well because its design required the analysis of relatively short sound clips, of only a few seconds' duration. This contrasts with musical notation, which is virtually unlimited in its ability to render long aural sequences. Students using musical notation preferred birds with varied and often extensive repertoires, such as the song sparrow. Spectrograms, instead, accommodated only vocalizations that were short and preferably repetitive, such as calls or songs of species like the chaffinch, whose variations were concise, stereotypical, and

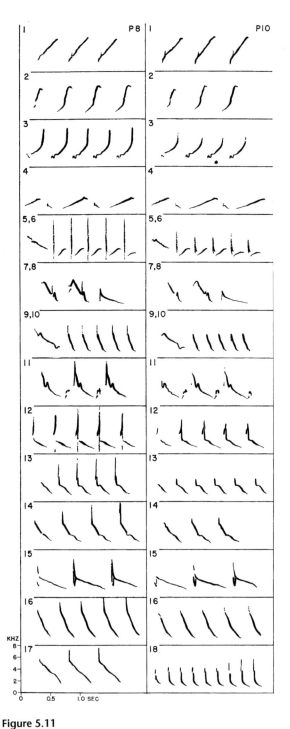

Figure 5.11

Sonagrams enabling the synoptic comparison of songs of two individual birds in one location.

Source: R. E. Lemon and A. Herzog, "The Vocal Behavior of Cardinals and Pyrrhuloxias in Texas," *Condor* 71 (1) (1969): 6. Reproduced with kind permission of the American Ornithological Society.

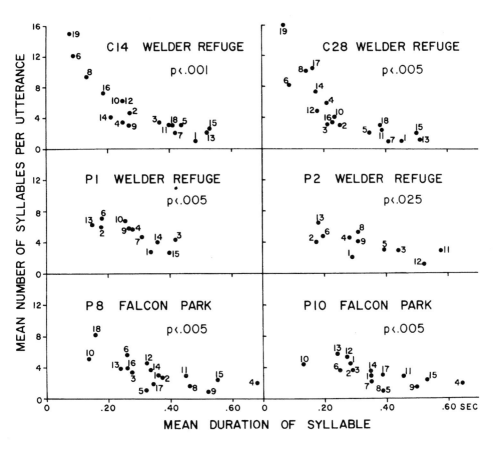

Figure 5.12

One further cascade: graph representing the statistical relation between the duration of syllables and the mean number of repetitions of the syllables for each song, for six separate individuals in two different localities.

Source: R. E. Lemon and A. Herzog, "The Vocal Behavior of Cardinals and Pyrrhuloxias in Texas," *Condor* 71 (1) (1969): 8. Reproduced with kind permission of the American Ornithological Society.

thus easy to understand "at a glance." The differences between these short song samples were quantified by attending primarily to the distribution of sound over the vertical frequency range, rather than to its horizontal unfolding in time.

Once printed and categorized, hundreds of spectrograms were measured and the extracted information punctualized into a matrix, table, or graph. More than images, numbers can be powerfully mobilized as proof in

Latour's model, because they make it increasingly costly to dissent. As the rift between researchers at Cornell and Cambridge regarding spectrographic retouching demonstrates, single images themselves could be disqualified or disbelieved based on the way they had been produced or interpreted. They could even be countered with another image. To argue against the graph as an abstraction of many individual sound events, however, requires one to muster at least an equal mass of sound events, abstracted from the field site and cascading with equal speed and efficiency from recording, to spectrogram, to numbers, to graph.

Pragmatic Conversions

The sound spectrograph effectively exposed the perceptual limits of human hearing. Birdsongs looked more intricate on paper than they had seemed to the ear. The possibilities that looking at birdsong offered were gratefully incorporated in a new analytic tradition, which found in it a key to uncover mechanisms of song variation and by extension of learning and speciation. The spectrograph did not, however, dispense entirely with the expert ear or specific sonic skills in the field of bioacoustics: the listener got displaced, but not replaced. The point of this analysis is not to extend or restore a putative hierarchy of the senses into the bioacoustics laboratory, but to use it as an entry point to further our understanding of the complex dynamic between an instrument, representation, and the researcher's body. What role did birdsong biologists negotiate then for embodied listening alongside the conventional spectrographic visualization?

With the sound spectrogram, birdsong biologists subscribed to a desire for an almost synesthetic interconversion of audible vibrations into visible tracings, which has been a persistent concern for instrument developers and acousticians since the early nineteenth century. In contrast to musical scores or other graphic schemes, which seemed to represent acoustic phenomena in an arbitrary fashion, automatic sound images were regarded as a natural analog to sound itself. Underlying these images was a holistic assumption that sound would be able to write itself, and that, eventually, traces could stand in for actual sounds without a loss of information. This motif of synesthetic inscriptions was embodied by the phonautograph and the phonograph, but it also surfaced in the Bell Labs experimenters' projection of a visible language for the deaf, as well as in the graphic modifications of sonagrams by ornithologists in Cambridge and elsewhere. Like Thorpe's diagrammatic notations or Marshall's logarithmically scaled spectrograms, they aspired to a mechanical and universally legible codification of natural sounds. However, such correspondence was never easy to achieve in practice: the conversion of audible sound into such visual formats introduced a new and different resolution in which sounds could be considered by frequency and temporal distributions of acoustic energy. But it also confronted researchers with a loss of information in acoustic energy *as sound*, both to the human and the animal listener.

It is here that sonic skills evidently played a significant role. Listening was mobilized at moments when the limits of the spectrographic image itself became obvious and visual information failed to stand in fully for auditory perception. This was most clearly the case when the interpretation of visual traces was adjusted, classified, or corrected based on auditory impressions. Here, the perspective shifted back and forth between a sound's physical measurement and its actual perceptive qualities.[27] The difficulty in establishing a consistent correspondence between sound and trace, and in distinguishing pattern from noise, forced researchers to continue to rely on their own embodied perception—if only to assess the degree of correspondence between these formats. In such cases, hearing was mobilized as a complementary "sensuous technology" that helped to make sense of acoustic phenomena (Roberts 1995, 507).

Indeed, although spectrograph users had publicly pitted the "objectivity" of the spectrograph against the "subjectivity" of earlier methods, this does not mean that these were irreconcilable epistemic positions per se, as researchers like Thorpe routinely integrated both aspects in their practices. At the same time, the precise conditions under which trained listening was allowed to supplement the spectrograph were carefully delineated. If trained judgment was invoked, it was as a pragmatic solution, more than as an epistemic commitment to the ear. Listening did not disqualify the spectrographic image, but rather supported and sustained its operation—it provided a way to repair temporary losses of information in what otherwise counted as a successful conversion of sound into image. For instance, while visual patterning was useful in drawing up morphological classifications, categorizing sounds by their graphic shape, it was not always the most efficient approach, since spectrographing hundreds of recordings was a very time-consuming task. It was at this point that auditory impressions were mobilized, with classifications initiated on the basis of aural impressions before spectrographing began. Such exploratory or supportive work could not easily be codified or carried out mechanically, because it was often grounded in the somatic experience of field recording. In turn, although sound spectrograms were deployed to aid and guide observations in the field, they did not do away with actual, sustained experience in the field. These instances of listening were, however, not facilitated by the spectrogram. Sound spectrograms were, after all, particularly useful to bioacousticians in producing authoritative inscriptions, precisely because they congealed auditory information in a way that made it difficult to be reopened to the embodied (and likely individual) experiences of the reader. These pragmatic considerations in which practices of trained listening and sonic skills were integrated in bioacoustics demonstrate interactions between researchers and instruments that are often different, sometimes richer, and sometimes simply more efficient than could be obtained by the mechanical image alone.

6 Conclusion

Bioacoustic Techniques

Noise matters. Not just for bioacousticians and field recordists, but also for animals. We tend to think of noise as nuisance, extraneous energy, or a disrupting ambience that impresses itself on the observer. Noise interferes with our ability to discover and interpret the more interesting signal. No wonder bioacousticians went to such lengths to suppress it, exclude it from their laboratories, filter it from their recordings, or erase it from their spectrograms—or that field recordists continue to flee densely populated areas in search of a few remaining spots that are still isolated from advancing waves of anthropogenic sound. But noise is also inherent in the more or less natural environments in which humans and animals manage their lives.

As such, noise is not just nuisance but part of the signal. In 2003, a brief paper in *Nature* impressed this realization on ecologists and a general audience. Dutch ecologist Hans Slabbekoorn and his colleagues at Leiden University reported a remarkable behavioral change in city birds (Slabbekoorn and Peet 2003). While studying the effect of noise pollution on vocal communication, they had found that great tits living in noisy urban locations generally sang at higher frequencies than their conspecifics in forested, quieter environments. Birds, the ecologists proposed, might well be adjusting their song to the acoustic conditions of their environment. Subsequent studies suggested that a sudden rise (sudden on the evolutionary time scale of the species, that is) in low-pitched anthropogenic (mostly traffic) noise was likely to be constraining communication between birds. Such ambient noises mask vital signals for a bird whose songs and calls are the primary means by which it defends its territory, attracts a mate, signals danger. Rising noise levels thus pose a potentially dramatic evolutionary disadvantage. Hence, just as humans would in noisy surroundings, in line

with the Lombard effect, birds have begun to vocalize at increased amplitude and at higher frequencies.[1]

The image of the small urban tit straining to sing above traffic noise
struck a chord, and the finding was widely picked up in the academic and
popular press. In part, this had to do with the bird's resilient and remark-
ably flexible adaptation to environmental changes. But the story also
powerfully drove home the realization that human and nonhuman soundscapes are not separate realms. More than forty years after Rachel Carson
(1962) warned of the prospect of a "silent spring," a chorus of birds growing
increasingly loud and shrill instead led back to human activities as its root
cause. The finding seemed to validate the long-standing concerns of field
recordists and conservationists that anthropogenic noises were compromis-
ing existing habitats and their soundscapes at an alarming rate. Since the
mid-1970s, some biologists had already begun to consider birdsong part
of a wider acoustic setting, examining the particular ways forests reflect,
deflect, and attenuate its vocalizations (Morton 1975).[2] Their work suggested that each species might occupy an acoustic "window" of optimal
frequencies within which its signals transmitted best (Marten and Marler 1977; Nelson and Marler 1990). But it was soundscape ecologists who
included the human in this acoustic equation. In 1977, for instance, the
composer R. Murray Schafer coined the phrase "acoustic ecology." He identified anthropogenic noise as the main driver of a shift from preindustrial
high-fidelity to postindustrial low-fidelity soundscapes. In 1993, musician
and soundscape ecologist Bernie Krause appropriated Morton's notion of
the "sound window" in his hypothesis that the vocalizations of each animal species occupy particular frequency bands or "spectral niches." Under
the influence of advancing urbanization, Krause warned, such niches were
increasingly at risk of being masked by a constant low-frequency sound. As
such, human activity threatened to disrupt not only a harmonious soundscape but an entire ecosystem.

It is perhaps no coincidence that anthropogenic noise should have
attracted the attention of the composer first, well before it was picked up
by the bioacoustician. That ambient noise would become a topic in the
study of birdsong at all was not self-evident, given the trajectories I have
described in this book. While some field recordists developed an aesthetic
interest in ambient sounds and natural backgrounds, bioacousticians have
historically become accustomed to regarding such ambience as an acoustic
interference and an aesthetic and analytic nuisance and their concern has
led to many of the tools, techniques, and routines used to record and analyze birdsong. Over time, directional recorders and parabolic microphones,

spectrographic adjustment, and soundproof laboratories and aviaries permitted the researcher to control the acoustics of the "natural environment" in a way akin to the sound engineer. Even when Slabbekoorn and Peet set out to track the influence of noise on the great tit's city serenades, they first separated its vocalizations from the environment on different tracks, by sampling the bird's vocalizations with a highly directional microphone and at the same time but separately recording its ambient surroundings with an omnidirectional microphone.

The pervasiveness of sterilized listening illustrates how technologies of recording and listening melded together in an "observational machinery": a machinery allowing not just field recordists themselves, but also their readers in scientific journals, to listen to events in the field. That machinery— really a chain of translations in the material existence of recorded sounds that links the naturalist listening in the field to the analyst in the laboratory, and back—was assembled by academic researchers, sound archivists, public broadcasters, and a diverse array of naturalists, birdwatchers, and field recordists with the help of engineers, musicians, and teachers. Drawing on these various backgrounds, they pieced together existing technologies of sound recording and listening practices and invented new ways of talking, exchanging, and archiving sounds. I have traced those material, social, discursive, and often sensuous technologies through phases of challenge and reconsideration. But once they stuck, they gradually sank in to become accepted techniques for making sense of birds' behavior. As such, the history of this machinery is as much a history of the kinds of knowledge that it *did* condition as that which it *did not*.

In tracing this process, this book's scope has necessarily been limited to a number of key settings, actors, and technologies, in a short time span of innovation, when novelties began to congeal into routines and users' interests crystallized in material forms. Doing so has revealed a set of basic bioacoustic techniques in the making. For sure, such techniques continued to develop well after the periods of innovation I consider here. In many ways, digital audio has remediated the analog media and the practices organized around them. iPods and mobile software applications now allow for easy playback and instant verification of identifications in the field. Scientists, field recordists, and birdwatchers exchange and discuss recordings directly on online platforms. And they continue to worry about sound fidelity, especially now that the MP3 format is routinely used to compress sound data for storage and circulation. Digital (audio) technologies have enabled animal vocalizations to be recorded, analyzed, represented, and shared faster and more widely. But they do not divorce themselves from the modus operandi of analog techniques, such as the directional microphone, the sound archive, and some form of transcription. Examining the local conditions under which these techniques' practices and meanings initially settled brings into relief attendant shifts that have shaped the practice of bioacoustics and continue to do so, in the professionalization of the field-worker, in the legitimacy of embodied observation, in the validation of scientific evidence, and thereby in the relationships between listener, animal, and its environment.

As a history of bioacoustic techniques—as opposed to the discipline of bioacoustics—this story is only one of many threads. Its focus on British and American ethology, through key figures at Cornell and Cambridge universities, leaves other geographic contexts, material conditions, and intellectual traditions unaccounted for. Bioacousticians have established sound archives worldwide, have affiliated with such disciplines as physiology, neuroscience, or linguistics, and have singled out a wide range of fish, insects, amphibians, mammals, and cetaceans as research objects. The formation of this expansive, interdisciplinary, and international field remains to be excavated further, especially since funding structures, intellectual and practical alliances (both successful and failed), and the various roles of military science and commercial interests will have prompted each of these specializations to develop at varying speeds. Indeed, the striking diversity of bioacoustics led conservationist Max Nicholson in a public address to field recordists and sound archivists in 1974 to lament its "terribly fragmented" state (Nicholson 1974). In his estimation, the field lacked cohesion, identity, and an awareness of the potential demand for its findings. At a time of publicly funded research programs, those issues had left it in a Cinderella relationship with stronger and richer research subjects. Nicholson's pessimistic account may have been inspired by his familiarity with birdsong research in particular, which was a topic close to his heart, but also one that gestured less ostensibly to Cold War sensibilities than other bioacoustic problems at the time (such as, for instance, underwater communication). But even if bioacousticians struggled to establish a shared definition of research objectives, by the mid-1970s, they did have a solid electroacoustic orientation to those objectives in place. This was due not least to students of birdsong, whose pioneering practices of recording, analyzing, and presenting acoustic phenomena were widely adopted across the field's many specializations. In that sense, the problems facing birdsong biologists exemplify (though are not always exactly congruous with) the way that bioacousticians as a whole began to listen to the natural world.

Recordings as Scientific Objects

As a history of the sound technologies put in place to record, organize, and represent sound scientifically, this book complements a growing list of objects and technologies that populate social studies of science and technology—from cinematic techniques over microscopes to colorful brain scans and online databases. One of the questions animating this book is how technologies of sound recording and listening are to be understood in relation to other such (predominantly textual, numerical, or visual) inscriptive technologies. In some ways, at least, it has become clear that they are not so different at all. Sound recording opens the notion of "inscription device" to include a whole class of other technologies that transform the sensory world into reproducible and transportable traces. Recording has afforded users an unprecedented control over the spatial and temporal dimensions of acoustic phenomena. Complex soundscapes could be sampled and sterilized, acoustic events repeated endlessly, slowed down infinitely, settled in spectrographic and numerical forms. The resemblance between recording and other inscriptive processes is not coincidental, as it allows sounds, like other sensory phenomena, to be counted rather than listened to. This has offered powerful analytic benefits, but it has ultimately also limited the interpretation of fleeting acoustic phenomena to black-and-white fact. Considering scientific sound recordings through this lens alone, however, obscures as much as it reveals. Even when they resembled other kinds of inscriptions, sound recordings were part of different technological, cultural, economic, and social contexts, which continued to influence their use. To consider scientific recordings only within a cascade of inscriptions disregards the material and cultural existence that has made them amenable to such frequent reuse and reinterpretation.

By considering sound recordings as scientific objects, then, this book brings together two important themes in social studies of science and technology: the versatility of scientific objects and the entanglement of knowledge with the bodily senses. First, thinking with sound recordings has helped bring to light the potential versatility of scientific objects and the ways they move in and out of scientific and technical communities. Even as sound recordings were ultimately processed as scientific objects, they remained cultural artifacts with commercial, artistic, educational, and sometimes even military appeal. As recordings were imported from these domains into scientific investigation, new concepts, technologies, people, and cultural, economic, and technical resources often followed in their tracks. And recordings often traveled in the other direction. The French composer Messiaen, for instance, transcribed most of the species in his famous *Oiseaux Exotiques* from gramophone records such as *American Bird Songs*, which ornithologists at the Cornell Laboratory of Ornithology published in 1942 with the conservation-oriented publisher Comstock on the basis of their field recordings (Dingle and Simeone 2007, 117). Recordings originally produced for scientific study were routinely repackaged in new forms, as commodity items, forms of entertainment, and instruments of popular instruction. Although this traffic has rendered the boundaries of scientific investigation porous, scientists have often tried to use this porosity to their own advantage, by knitting their own ambitions in with those of other professional and amateur communities. Such informal affiliations and alliances are crucial in understanding how sound recording came to structure what biologists heard because they helped to carve out niches where field recording flourished—even when a secure disciplinary or institutional footing was lacking—and thus ensured researchers' access to valuable instruments, a steady supply of data, or the help of volunteer collaborators. And because in relying on other professional communities—whether they were musicians, public broadcasters, or communication engineers—scientists also unmistakably adopted some of their ways of thinking and talking about sound. How scientists hear birdsong today is thus shaped in part by the histories of those exchanges.

Second, the history of scientific sound recording draws attention to the embodied practices that infuse scientific knowledge production and reminds us that researchers' situated and tacit knowing involves more bodily senses than their "eyes and hands" (Latour 1986). For sure, when field recordists began to embrace electroacoustic concepts and instruments, the trained ear seemed to retreat from being the primary analytic instrument to become a mere technical precondition. Advanced sound technologies exposed the human ear as weak, unable to pick up on auditory richness beyond normal hearing thresholds. Systematic field surveyors were advised to take particular care in selecting observers, for fear that a lack of training or hearing loss—a problem with people over forty—would significantly alter results (Emlen and DeJong 1991). More recently, electronic listening devices partly automated that work—but only partly, as expert listeners continued to be deployed for checking and interpreting these records. Listening and other corporeal skills continued to be epistemically significant in several ways. In communicating sound, the body itself often served as an effective metaphor. The 1992 edition of Hamlyn's *Birds of Britain and Europe*, for instance, utilized around 120 adjectives to convey the quality of bird sound, many of which drew on bodily experiences. The sound of the sandwich tern, for example, was described as "amalgam being pressed into a tooth." Such skills also continued to play a role in interaction with the instrument, to complement, monitor, and fine-tune processes of recording and inscription. Certainly, from the 1930s onward recordists needed skills in electrical engineering and acoustics, more than musical literacy. But they also needed a significant degree of auditory skill to ensure that technical defects would not alter the quality of the recording. Magnetic tape recorders had to be monitored for fluctuating tape speeds to prevent the creation of artifacts in the recording that would make it useless for analysis. The practiced ear also remained a valuable asset—not always for charting complete acoustic (or musical) structures from scratch, but certainly in tasks of recognition, diagnosis, and categorization. Listening facilitated the transformation of records into convincing inscriptions, and sometimes even their interpretation. Finally, field recordists continued to listen in the field, which they navigated with their own ears to locate, identify, and describe bird vocalizations.

Both themes—the dynamism of scientific objects' circulation and reception and the continued relevance of embodied skills—are intimately related. Listening may have been transformed into an individual, tacit intuition, whose personal and variable character made it unfit for producing objective knowledge. But particularly in a field environment, listening was also a mode of attention that had to be practiced. Against the background of field biology's professionalization, the question of who could listen authoritatively and in what way was negotiated largely through sound recordings. This was because the wide circulation of sound recordings activated very different sets of locally constructed skills in producing them. In turn, these recordings acted as concrete technologies of education and attention, training ears and sensibilities of fieldworkers and others. This is best illustrated by the Cornell Library of Natural Sounds, where recordings fueled an intricate economy of exchange, reward, and accountability that disciplined field recordists' listening: recordings set examples for good recording, were used to train outgoing fieldworkers, served to check their field skill and reliability upon return, and could lead to commercial publication by way of reward. This notion, and a deeper understanding of the multiplicity of "scientific data," hold valuable lessons for discussions within the context of commercialization in scientific research, citizen science, and the governance of the knowledge commons. Within its circular economy, the CLNS freely drew on recordings to distinguish volunteer scientists from unpaid laborers and infuse them with the skill, knowledge, and habitus of the "scientific recordist."

Above all, recordings helped to standardize a discourse. Like their visual counterparts, the field guides, these recordings established a shared register for describing individual birds encountered in the field. And they established a template for thinking of birds and recordings in acoustic terms: complex and variable song patterns were usually represented by a few, brief segments of "typical" vocalizations against a sterile, static, and unidentifiable background. Regardless of listening ability, birdsong could not be described authoritatively without a stable perceptual framework that enabled field observers to judge whether what they heard in the field was the sound of a bird, whether it was new or familiar, of which species and what behavior it displayed, and whether it was potentially relevant or not. Precisely because all recordists—academic, commercial, and amateur— remained dependent on various monitory, exploratory, and diagnostic listening abilities in their fieldwork, they required what John Law and Michael Lynch (1988) have termed a "table of possibilities," a mental or concrete taxonomic array by which ornithological listeners may organize their perception in the field and bring their own tacit, idiosyncratic apprehensions of a sound in accordance with a standardized aural discourse. For this aural enskillment, listeners relied not only on the experience and field craft that they acquired over time and by themselves, but also on field guides and recordings, spectrograms, and notations. Because birdsong biology was reliant on practiced listeners in the field, the organization of scientific sound recording has always necessarily been interwoven with questions of pedagogy and instruction.

This is an important point, because the need for field listeners' aural enskillment dictated a constant need for reconversion: from the *visual* realm of the spectrographic trace to that of the observers' own *aural* experience, back in the field. This tension between analysis and pedagogy has been of continuous concern to birdsong biologists since 1900. It is for this reason that around 1915, musical diagrams were complemented with graphic notations that were claimed to be more easily readable and memorizable in fieldwork. And it is for the same reason that a few biologists and professional musicians in 1965 collaborated to mend the sound spectrograph, in turn, with musical notation. But this relationship between analysis and didactics was tenuous. Records were supposed to strip sound of any ambiguities related to its acoustic experience and at the same time to convey a tacit, experiential knowledge of those sounds in the field. Sound was both to have an objective existence and to be grasped subjectively. Now that techniques of visualization are no longer the privilege of advanced biological laboratories, and digital software packages and portable computers make spectrographic analysis easier than ever, experienced ornithologists claim a solution to this problem. Making sound spectrograms allows one, in the words of the prominent New York ornithologist Don Kroodsma (2007, 409), to "watch the songs dance across your computer monitor in the great out-of-doors as you listen to birds there. Listen as you see, and you will hear a different world singing to you." Part scientific textbook, part manual, part personal journey, Kroodsma's venture into the art and science of listening to birdsong promises insight into the private lives of birds. It is possible that one can circumvent one's "pathetic ears" by learning to listen as if a sound were a spectrogram, as William H. Thorpe had hoped (Kroodsma 2007, 1–2). But the abundance of teaching tapes and software applications on the market suggests that listening continues to be key to investigating birds in the field.

The ever-present tension between encapsulating and mediating sound knowledge, between listening for closure and listening to provoke, also explains why recording formats continued to coexist as they did in the period studied here. New types of inscriptions did not simply obliterate one another; instead, they accumulated, proliferated, and extended into local complexes of technical and social practice. Around 1965, bird vocalizations typically existed in multiple, duplicate forms. Scientists as well as amateur and professional recordists routinely captured, edited, archived, or exchanged a vocalization on magnetic tape, which was cheap, versatile, and easy to montage, but also vulnerable to "tinkering" or even erasure. The same bird vocalization could be converted to the robust medium of a gramophone record, a medium of choice for pedagogical purposes and popular distribution. Or it could be found scribbled down as a graphic diagram in field notes or laboratory reflections for the recordist's own reference, or even be rendered as a musical score to discuss with expert listeners. Finally, it would exist as a spectrogram whose expense and technically complex instrumentation effectively privileged the professional biologist. All these media captured and codified sound in different ways, followed different conventions, appealed to different audiences and functions, and were invested with different degrees of scientific authority. While some media adhered more closely to an ideal of the graphic coordinate and measurable object, others shared only some of the characteristics that seem to make an authoritative inscription. Those inscriptions did, however, possess other perceptual, semiotic, and epistemic operationalities that were important in producing a body of nature listeners.

Considering these media, materials, and formats within the same register of "recording technologies" shows how sound inscriptions do not just follow one vector of successive transformation into a single authoritative data point. Instead, these vectors feed back into a cyclic, self-reinforcing loop, where recordings are continuously converted, between different media, between social and professional contexts, and between different sensory modalities. Sound recordings were never *only* scientific objects. Recognizing this range of possible roles, this book has shown that the production and legitimization of scientific records have almost invariably also depended on their distribution as objects of popular instruction, amazement, and joy.

Notes

1 Eavesdropping in the Wild

1. "Voices of Southern Birds Recorded for Movie Films," 1935, 316. On suspected causes of the disappearance of the ivory-billed woodpecker, see Beyer 1900.

2. Quoted in Walters and Crist 2005. Also see Barrow 2011.

3. See, for instance, Dalton 2005; Fitzpatrick et al. 2006; Hardy 1975; Hill et al. 2006; Jones, Troy, and Pomara 2007. For an analysis of the visual evidence and its role in this controversy, see Lynch 2011 and Winn 2009.

4. This comparison is made in Kroodsma et al. 1996.

5. Thompson's (2002) work on the early twentieth-century history of architectural acoustics has been particularly influential in showing how scientific study and technical manipulations of sound through electroacoustic instruments and sound-absorbing building materials helped to redefine conceptions of architectural space and culminated in a distinctively modern soundscape. More recently, cultural historians have traced the relations between specific acoustic knowledge and the historical and cultural contexts within which investigators and their collaborators operated, such as noise control (Bijsterveld 2008), instrument manufacturing (Jackson 2006; Pantalony 2009), communication engineering (Mills 2010, 2018; Sterne 2012), music making and composition (Hui 2013; Hui, Kursell, and Jackson 2013), theater architecture (Tkaczyk 2014), and applied research and engineering (Wittje 2016). This work differs from conventional histories of scientific acoustics' longue-durée perspective of key investigators (Hunt 1978; Beyer 1999) or focus on the role of music and acoustics in the making of early modern science (Erlmann 2010; Pesic 2014).

6. For an analysis of listening practices in the concert hall, see Johnson 1995. Such work draws on both historical (see, for instance, Lachmund 1999, Kursell 2003, and Volmar 2015) and ethnographic analysis of listening practices (Mody 2005; Rice 2008). Pinch and Bijsterveld (2012) have emphasized the importance of sonic skills

(such as listening, but also producing, recording, storing, and analyzing sounds). Drawing on detailed empirical studies of medical professionals listening to hospital equipment and patients' bodies (Harris and Van Drie 2015), car mechanics and engineers listening to engines (Krebs 2012), astrophysicists and high-energy physicists listening to auditory displays of stellar oscillations or particle collisions (Supper 2012), or indeed ornithologists (Bruyninckx 2012), Supper and Bijsterveld (2015) have further parsed listening into distinctive modes of attention. Investigators' reasons for listening and ways of listening differ, they argue, which has repercussions for the knowledge claims that can be made and the status that they afford the auditors.

7. Sterne (2003) has coined the term *audile technique* to describe the nineteenth-century reconstruction of listening in science, technology, medicine, and industry into a distinctively modern way of knowing and interacting with the world. Listening, according to Sterne, became a technical skill that could be developed, used to instrumental ends, and afford to listeners a great deal of currency (Sterne 2003, 94–95).

8. See Daston and Lunbeck 2011, 1, 8; Fleck 1979.

9. In the history of science, see, for instance, Smith 2004, Lawrence and Shapin 1998, and Roberts 1995. In the sociology of science, see Myers 2008 and Knorr Cetina 1999. In the history of technology and environment, see Parr 2010 and Murphy 2006.

10. See MacDonald 2002 and Toogood 2011 specifically on observation in ornithology, and Vetter 2011a on lay observation in the field sciences. On the organization of observation in the field, see Law and Lynch 1988 and Ellis 2011.

11. There is a rich literature on the history of sound recording that focuses on the one hand on the production of meaning, surrounding its earliest technologies, and on the other, on sound recording in the record and entertainment industry (Morton 2004; Gelatt 1955; Read and Welch 1976; Gitelman 2008; Schmidt Horning 2013; Katz 2004; Taylor 2014; Day 2002; Morton 2000).

12. The history of sound recording has been documented extensively for the field of cultural anthropology and comparative musicology. See, for example, Ames 2003; Stangl 2000; Sterne 2003; Brady 1999; Nettl and Bohlman 1991; Rehding 2000. On recording in the history of linguistics, see Hochman 2010; Kaplan 2013; Lange 2015. Recordings have also been used in psychophysics and experiments on the physiology and psychology of hearing; see Hui 2013 and Kursell 2012. For work on the historical use of sound recordings in psychology and animal behavior, see Thomas 2005, Radick 2007, and Burnett 2012.

13. In his accomplished history of post-Darwinian primatology, for example, Gregory Radick (2007) traces the intellectual circumstances that have led phonographic

recordings of primate vocalizations to recharge (in several instances) a debate on the concept of the animal mind and the evolution of animal and human language.

14. Daston and Galison's (2007) term *mechanical objectivity* describes a set of virtues that was coincident with the emergence of mechanical reproduction in the late nineteenth and twentieth centuries, including sound recording technologies such as the phonograph.

15. For a comprehensive overview of studies of images and visualization, see Burri and Dumit 2008. A recent volume by Coopmans et al. (2014) revisits a volume edited by Michael Lynch and Steve Woolgar (1988), which serves as a cornerstone for this line of research.

16. See among others Kohler 2002a, 2002b, 2006; Kuklick and Kohler 1996; Vetter 2011a; Lachmund 2013; De Bont 2015.

17. This may also be gleaned from recent primers on (early) bioacoustics by Radick (2007) and Burnett (2012). Their stories intersect in numerous ways with my own, with historical actors drawing on similar resources—personal, technical, and intellectual.

18. On the culture of technological hobby and ham radio in particular, see Haring 2007. On amateur radio, see Douglas 1992.

19. On sound hunting as a technical hobby, see Bijsterveld 2004 and Massinon 2016.

20. See, for instance, LaFollette 2008, 2013; Kirby 2011; Davies 2000; Mitman 1999.

21. North American, British, and Central European ornithologists are strongly represented in the latest overview of scientific ornithology since Darwin (Birkhead, Wimpenny, and Montgomerie 2014). For the history of US and Canadian ornithology, see Davis and Jackson 1995; Ainley 1985; Barrow 1998, 2000; Allen 1951. On ornithology in Britain, see Allen 1976; Davis 2011; MacDonald 2002. On ornithology in Germany, see De Bont 2009; Lachmund 2015; Haffer 2008, 2007. Recently, this traditional focus has been expanded in other geographic regions (see Jacobs 2016).

22. This is indicated by the distribution of contributors to the important volume edited and published by René-Guy Busnel for the inaugural meeting of the International Committee of Bioacoustics. Among the twenty-four contributors, six were associated with American research institutes, three were located in England, three in Germany, one each in the USSR and Finland, and an overwhelming ten were affiliated with French research institutes, including a large delegation from the Laboratoire de Physiologie Acoustique in Jouy-en-Josas.

23. Radio engineer Carl Weismann, for example, began recording wild bird sounds for Danish Radio near Copenhagen in 1934, while Gunnar Lekander and Sture

Palmér pioneered bird sound recording in Sweden for Radiotjänst AB (Swedish Radio) in 1936, and the Japanese ornithologist Tsuruhiko Kabaya published bird sound recordings in 1954 (interview with Sture Palmér by Patrick J. Sellar, March 30, 1983, British Library C90/03/01; interview with Carl Weismann by Patrick J. Sellar, March 28, 1983, British Library C90/02/01). In France, Claude Chappuis recorded avian sounds for his own scientific use, but he published an album of bird sounds and deposited his recordings with the wildlife section of the British Library Sound Archive (Copeland, Boswall, and Petts 1988).

24. The Macaulay Library was formerly called the Library of Natural Sounds. The British Library of Wildlife Sounds emerged as part of the British Recording Institute, which itself was built in part on an extensive collection of wildlife sounds by the British Broadcasting Corporation and the personal collection of recordist Ludwig Koch.

25. See Hughes 1983; Jacob 1999; Van der Vleuten and Kaijser 2006; Turchetti, Herran, and Boudia 2012.

26. I examined the journals *The Auk* (published 1884–present), *The Condor* (1899–present), *The Wilson Bulletin* (1889–1999, now *The Wilson Journal of Ornithology*), *Journal of Field Ornithology* (1930–present), *Ibis* (1859–present), *British Birds* (1907–present), *Journal für Ornithologie* (now *Journal of Ornithology*, published since 1853), *Animal Behavior* (1958–present), and *Behavior* (1947–present). Work published elsewhere was considered when it was referred to by key authors.

27. I selected these journals by following the publication trajectories of researchers and recordists, making no distinction between full academic papers and shorter contributions under headings such as communications, reviews of books and records, editorials, or "notes"—in fact, these rubrics proved particularly relevant for the spread of methods and ideas through the community.

28. See the references for a list of archival sources consulted and their abbreviations used in citations.

29. For a discussion of scientific personae in scientific texts, see Daston and Sibum 2003.

2   Scientific Scores and Musical Ears

1. The literary genre of the nature essay, an invention of John Burroughs in the 1860s, became wildly popular in the 1890s, especially in the United States. Following in the footsteps of Thoreau, Whitman, and Burroughs, nature writers wrote of excursions to fields and woods and about birds and wildlife in people's own farms and yards, thus familiarizing an increasingly urbanized readership with the natural landscape (see Lutts 1990). Since the 1880s, notable nature writers such as Charles C. Abbott (1896) and Bradford Torrey (1885) had devoted considerable space to

exalted discussions of the birdsongs they observed on their ramblings. Mathews considered Torrey's novel to be of inestimable value to the study of bird music, more than all nature writings (Mathews 1904, x). Novelist Gene Stratton-Porter experimented with the genre in her novel *The Song of the Cardinal* (1903) as an extension of her naturalist work. The novel follows the fictional life of a cardinal, and was reputedly written in response to sport hunting. Her naturalist studies found a reflection, among others, in her book *Music of the Wild: With Reproductions of the Performers, Their Instruments and Festival Halls* (1910). Divided into sections such as "The Chorus of the Forest" and "Music of the Marsh," it calls on the reader to appreciate nature's music.

2. The use of musical notations to record birdsongs goes back at least to the notations of the Jesuit polymath Athanasius Kircher, who famously transcribed a nightingale song along with the voices of common fowl in his *Misurgia universalis* around 1650. This found an echo in the recorder notations collected in *The Bird Fancyer's Delight* of 1717, which were designed to teach birds to sing. For a brief survey of birdsong studies before 1850, see Rothenberg 2005, 19–22. However, Mathews took his cue not from these precedents but from two publications in the 1890s. *Wood Notes Wild: Notations of Bird Music* by the music teacher Simeon Pease Cheney, posthumously published in 1892, used simple Western musical scores to document birdsongs (while also hinting at the musicality even of inanimate things). These notations were among the first to be collected systematically: having approached two hundred ornithologists and librarians in the United States and Europe, Cheney found only a few scattered antecedents to his craft, such as Hermann Landois's (1874) *Thierstimmen* and the German periodical *Gefiederte Welt*, which was published from 1872 onward (Cheney 1892, viii). In Britain, naturalist Charles A. Witchell documented his extensive study on the evolution of birdsong in an appendix with simple musical notations for three common birds (Witchell 1896).

3. At least in form, Mathews's notation presents an intriguing similarity to a graphic notation that Stravinsky doodled at the request of director Robert Craft to "draw his own music." The notation is mentioned in Craft and Stravinsky 1952, 120, but it can also be admired on the covers of the journal *Perspectives of Music*. Musicologists tend to situate the age of revolution in musical notation in the postwar era, with experiments by John Cage and Earle Brown. The first half of the twentieth century, however, witnessed its own string of innovations in musical notation that originated not in the world of Western music but in scientific fields such as experimental psychology, ornithology, and comparative musicology. For an overview of graphic experimentation in musical scores, see among others Pryer 2002; Randel 2003; Sauer 2009; Taruskin 2005.

4. For a detailed portrait of such naturalists and an analysis of the kind of observations they collected, see Crist 1996. The term *civic realm* draws on Lynn Nyhart's (1998) analysis of Karl Möbius's scientific activity in the context of the civic culture of the German *Bürgertum* in the late nineteenth century.

5. See, for instance, Lachmund 1999, which shows how the codification and interpretation of lung sounds by means of auscultation remained subject to controversy throughout the mid-nineteenth century. French- and German-speaking medicine's attempts to standardize perception by various means were clearly shaped by, and themselves shaped, the local context in which these attempts took place. In his study of Feynman diagrams, David Kaiser (2005a, 2005b) has shown that the diagrams, along with the calculation techniques they enabled, spread through personal contacts between postdocs, but were also appropriated within their local contexts for new purposes and imbued with new meanings. For further discussion of the various ways instruments and tools bring together communities, see Clarke and Fujimura 1992 and Mody 2011.

6. See, for instance, Collins 2001; Kaiser 2005a, 2005b; Kohler 1994; Warwick 2003.

7. I take the notion of paper tools from Klein 2001. It highlights the role of diagrams not only in illustrating and expressing knowledge post hoc, but—like a laboratory tool—also helping to mold it. As such, the term prompts us to ask how the syntactic and semantic affordances of such tools meet particular epistemic needs.

8. For a history of these practices, see Barrow 1998; Bircham 2007; Burkhardt 2005; Mearns and Mearns 1998.

9. For instance by Burkhardt (2005), Haffer (2001, 2008), and Junker (2003).

10. See Doughty 1975; Kohler 2006; Lutts 1990; Orr 1992; Ritvo 1989.

11. For an early and detailed exploration of the "back-to-nature" movement, see Schmitt [1969] 1990. On the motives of field collectors and their relationship with leisure and outdoor recreation, see Kohler 2006.

12. The societies were established in 1850, 1858, and 1883 respectively. The Deutsche Ornithologen-Gesellschaft appears to have been mainly a society for museum collectors (though its objectives were never so obviously stated) (Haffer 2007).

13. Quoted in Birkhead 2011, 205. For a close study of Ridgway as a representative of scientific ornithology in the late nineteenth century, see Lewis 2012.

14. Griscom's own experience may have been a misleading benchmark here. Reputedly, he had the advantage of unusually acute hearing. On military duty during World War I, he is said to have become something of an "exhibition piece" for army doctors interested in the sensitivity of his hearing. His musical memory and critical ear certainly served him well in assessing the subtleties of birdsong. See Peterson [1934] 1965, 599. The issue of the "human" or "personal equation" was experienced more widely in the sciences; for a great discussion of the "personal equation" in astronomy, see Schaffer 1988.

15. For an excellent history of the ornithological field guide, see Dunlap 2011. For a critical analysis of the genre's ideological implications, see Schaffner 2011. On professional discussions regarding the epistemic value of sight records at the American Ornithologists' Union, see Barrow 1998, 154–180. On the construction of observation as a modern technique in new British ornithology from the 1920s onward, see MacDonald 2002 and Toogood 2011. For an ethnographic account of visual skilling in naturalist training, see Ellis 2011.

16. "Wenn ich aber durch Feld und Wald wandre, bin ich nie alleine. Da verkündet mir ein Vöglein mit unnachahmlichen Jubel seiner Liebe Glück." Here and throughout, all translations are my own unless otherwise attributed.

17. In fact, Beach's compositions provide a remarkable predecessor to Olivier Messiaen who, as von Glahn (2013) notes, is celebrated for pioneering the use of accurate transcriptions in his compositions in the 1940s and 1950s.

18. Mathews, for his part, was highly appreciative of any compositional references to birdsong. His discussions are laced with musical quotes of well-known compositions that referenced specific birdsongs and stayed true to their theme, whether in Beethoven's Pastoral Symphony or in nursery melodies by his contemporaries. He refers for instance to *Cuckoo! Cherry-Tree* by the contemporary English musician Joseph S. Moorat. Musical notation allowed such popular tunes to be juxtaposed with the recordings of actual birds (Mathews 1904, 20–21).

19. For the relation between tone and color in music pedagogy, see, for instance, Batchellor 1913, 45.

20. The movement was strongly linked to aims for preservation and conservation and its advocates promoted basic natural history, such as plant identification or animal life histories, in nature or school gardens, as both a hands-on antidote to and a way of promoting the skills necessary to succeed in modern industrial life. By 1907, nature study had been incorporated in many school curricula. For detailed histories of the nature study movement in the United States, see Kohlstedt 2010 and Armitage 2009. For an early history of natural history teaching in Germany, in conjunction with moral instruction and show-and-tell or *Anschauungsunterricht* (and its methods of hands-on instruction outdoors), see Nyhart 2009.

21. Lectures on natural history subjects were a popular element of the Chautauqua circuit. A collection of promotional materials from the Redpath Chautauqua circuit that included Oldys's lectures shows several advertisements for descriptive lectures on wildlife, often accompanied by motion picture materials or whistled reproductions.

22. As with many other bird whistlers, his vocal caprioles were recorded and published on record by Victor in 1913.

23. For a discussion of bird mimics and phonographic whistlers as a popular form, combining entertainment with a conservation ethic, see Smith 2015.

24. Porcello is concerned with the linguistic practices of professional studio engineers, showing that their status as professionals was closely tied to their competent use of linguistic (as well as other, more tacitly embodied) resources.

25. "Vom naturwissenschaftlich-musikalischen Standpunkte." For a detailed discussion of Hoffmann's premises, see Kursell 2003.

26. For an accessible explanation of Darwin's take on the function of birdsong, see Birkhead 2011, 266–267. The importance of Darwin's work for the study of animal music, and conversely, the relevance of the study of birdsong to the origins of melody, were made explicit early on (see Kivy 1959, 42–48). Oldys's remark that his observations of hermit thrush music might seem revolutionary should be considered against the background of this debate, which was revived by a critical commentary by Spencer in 1890, and its continued discussion, for instance, in Wallaschek 1893 and Stumpf 1911.

27. A good example of this in terms of the study of birdlife is Royal Dixon's (1917, xii) comment: "This is the secret of all worth-while nature study. We must look upon a bird as we do upon a man—not merely to learn the Latin names of bones and muscles, but to study its disposition, character, emotions, and thought processes. In other words, we must treat a bird as a friend and not as a scientific specimen."

28. For an overview of Saunders's life and works, see Joseph Brauner's "In Memoriam" (1979).

29. On Moore's life and works, see Friedmann 1964.

30. "Die gefährlichste Möglichkeit der Täuschung, der auch der geübteste musikalische Beobachter immer wieder zum Opfer fällt: die Gedächtnisintervalle, d.h. die gewohnten Intervalle unserer Musik, die wir in die objektiv gegebenen Tonschritte hineinhören, auch wenn diese von jenen sehr erheblich abweichen."

31. "Möchten übrigens auch dem Studium des Vogelgesangs zugute kommen."

32. The same point was made by the pioneering anthropologist Jesse Fewkes in 1890, when he arranged for his phonograph recordings to be transcribed for analysis in musical notation, incidentally by another skilled amateur bird recordist, Simon Cheney. "Blind Tom" refers to an African American musical savant who lived in the second half of the nineteenth century. In this context, he exemplifies the exceptionally skilled listener. See Brady 1999 and also Cheney 1892, 49.

33. This was also the case in comparative anthropology. See particularly Hochman 2010.

34. Musicologist Rachel Mundy (2010) has explored the currency of metaphors of dissection in ethnomusicological work in the first decades of the twentieth century. She finds that in rendering their recordings of exotic music as "specimens" whose features could be exposed through sustained analysis, the authors draw on the legitimacy of the laboratory and traditions of morphological and anatomical study in the natural sciences. I have not found any overt reference to this metaphor in ornithology. Yet if the work and claims to authority of scientist-musicians may be understood to follow a similar strategy of authentication, by affiliating their methods with systematist and faunists' concern with meticulous measurement and description, the graphic notations presented here, which abandon the accuracy of individual parameters in favor of a holistic concern with "structure" and instant recognition, may have been more likely to affiliate their work with the methodologies of fieldwork advocated by Ludlow Griscom and his followers.

35. See, for instance, Strong 1918 or Wheeler and Nichols 1924.

36. See Latour 1986. The term *conscription device* was developed by sociologist Kathryn Henderson (1991) to accentuate how drawings (such as engineering plans) may serve as network-organizing devices, enabling participants to engage each other and enlisting their participation.

37. See also Klein 2001; Cambrosio, Jacobi, and Keating 1993; Rudwick 1976.

3   Staging Sterile Sound

1. The term *soundscape* was first coined by composer R. Murray Schafer (1977) to describe the sonic elements that define an acoustic environment, but the term has since been appropriated widely in the interdisciplinary field of sound studies. In recent years, its environmentalist distinction between natural sounds and anthropogenic noises has been problematized and stripped of its ideological undertones. It has been redefined not just to encompass the physical environment, but also as a way of perceiving that environment that is technologically, conceptually, and discursively constructed (Corbin 1998; Greene and Porcello 2005; Thompson 2002).

2. On the emergence of electroacoustics and its redefinition of signal and noise, see Wittje 2016.

3. There has, however, been significant diversity in the function, design, and appreciation of the laboratory since the nineteenth century.

4. On the history of the field as a place of scientific investigation, see Kuklick and Kohler 1996; Lachmund 2013; Vetter 2011b.

5. This is evident, for instance, in histories of noise abatement drawn up by Bijsterveld (2003) and Mansell (2017), or in cultural geographies of nature experience and preservation (Coates 2005; Matless 2005).

6. This thesis has not remained wholly unqualified; it has been pointed out that other branches of field research were less strongly impacted by the premises of laboratory work, and that other field stations were also always more than the "placeless laboratories" they might first appear (De Bont 2015).

7. For a comprehensive historical overview of Allen's work in research and teaching ornithology at Cornell University, see Butcher and McGowan 1995; Allen 1933; Harwood 1987.

8. For a discussion of the role of new media in attracting public interest in bird-watching, bird behavior, and bird protection, see Wachelder 2009.

9. The recordings were assembled in a confidential report to the Office of Scientific Research and Development. In subsequent years, the laboratory published the fruits of the labor. First, in 1948, *Voices of the Night*, an album of four 78 rpm discs that contained the vocalizations of twenty-six frogs and toads, was published with surprising success (Kellogg 1962a). In 1950, the researchers published another set of recordings as *Jungle Sounds*. On the jungle acoustics project, see Eyring 1946.

10. Elsewhere in Germany, too, phonograph records of singing birds were included in museum exhibitions. In 1933, the Stuttgart Natural History Museum played recordings of German native species made at the Stuttgart Zoo in collaboration with the Württemberg wireless company alongside its taxidermic exhibits ("Voices of Birds Record," 1933).

11. Interview with field recordist Carl Weismann by Patrick J. Sellar, March 28, 1983 (British Library C90/02/01).

12. Interview with field recordist Sture Palmér by Patrick J. Sellar, March 30, 1983 (British Library C90/03/01).

13. See, for instance, Johnson 1995 on this shift in British ornithology. For an overview of similar developments in German and American ornithology, see Haffer 2007 and Battalio 1998.

14. For more on Huxley's role and ambitions as a science popularizer, see Kevles 1992. For an analysis of Huxley's film and its objectives, see Mitman 1999, 76–78. For Huxley's role in the rise of British ethology, see Burkhardt 1992.

15. Julian Huxley declared birds to be "expressions" of the nation, in effect mapping wartime anxieties onto Britain's birds. See Huxley 1949 and MacDonald 2002. Michael Guida (2015) has noted that birdsong was mobilized for therapeutic effect in literature and broadcasting to soothe the impact of two world wars.

16. Elsewhere in Europe, radio engineers collaborated with naturalists to produce recordings for educational purposes. In Denmark, recording engineer Carl Weismann began recording wild birds in 1934. Copies of his work were distributed by Dansk Stadsradiofonien to schools. In 1937, Weismann rented a recording van and

began selling his work commercially, earning a distinction in 1955 as the only nature recordist to earn a spot on the British Top Twenty charts with the sales of his novelty record *The Singing Dogs*, which provided funds to finance further recording expeditions. In Sweden, private detective and amateur ornithologist Gunnar Lekander and professional recording engineer Sture Palmér began recording Swedish birdsongs for an educational record in 1936, as part-time employees of the Swedish radio broadcaster Radiotjänst AB. In 1938, Swedish Radio acquired its own mobile recording van with Telefunken disc cutters; it released sixty-five 78 rpm records in the next two decades (Copeland, Boswall, and Petts 1988, 6–7).

17. Albert R. Brand to P. B. Coffin, January 28, 1932, Brand Papers (CUL), 1:26.

18. Such surveys were also carried out elsewhere. Hans Slabbekoorn (2009) notes a similar wide-ranging ear-based survey of the great tit in Finland between 1947 and 1950. The research was repeated in 1981, revealing a dramatic shift in the ratio of song types.

19. Albert Brand must have been aware of these ethologists' ideas. During a trip around Europe in 1938, he and Allen met both Oskar Heinroth and Konrad Lorenz. Brand reported that "[Heinroth] has recently recorded birdsongs and is especially interested in working up individual species, recording their calls and songs completely. As he says, song can be gotten in Nature, but calls are different and more difficult. He suggests hand-reared birds and recording mainly from birds of the second captive generation" ("Bird Hearing Experiment," March 9, 1938, Brand Papers [CUL], 1:40).

20. "Es handelt sich hier also in keinem Falle um gekäfigte Tiere. Diese Art der Aufnahme verbürgt eine tadellose, ich möchte sagen, eine standortsgemässe Naturtreue."

21. For more on the art of whistling, see Tipp 2011a, 2011b, and Smith 2015. Likewise, well into the 1930s caged canaries and nightingales were being recorded in combination with musical arrangements of classic tunes such as *The Blue Danube*. See Bevis 2010; Birkhead 2003; Tipp 2011c.

22. "Derartige Aufnahmen nicht in Zoologischen Gärten, aber auch nicht im Zimmer oder im Aufnahmeraum einer Schallplattenfabrik gemacht werden konnten, sollten sie wirklich naturgetreu ausfallen."

23. See, for instance, Thompson 1995 and Sterne 2003.

24. See Vetter 2004 on science along the railroad, or Kohler 2006 on the connection between naturalist collecting, leisure, and mobility.

25. Interview with Carl Weismann by Patrick J. Sellar, March 28, 1983 (British Library C90/02/01); interview with Sture Palmér by Patrick J. Sellar, March 30, 1983 (British Library C90/03/01).

26. See histories of noise abatement drawn up by Bijsterveld (2001, 2003, 2008), Mansell (2017), and Thompson (2002).

27. Interview with Magnus Robb by the author, March 8, 2008; interview with Roger Boughton by the author, May 14, 2010.

28. The quotation is taken from Lynch 1985, 43. Also see Amann and Knorr Cetina 1988; Latour 1986, 2000; Lynch and Woolgar 1988.

4   Sampling Assets

1. Report *International Committee on Biological Acoustics*, n.d., CUL, Kellogg Papers, 3:30; see also Busnel 1963, ix.

2. Peter Paul Kellogg to Eric Simms, May 2, 1956, WAC, BBC Records, R46-365-2.

3. For the published proceedings of the ICBA's inaugural meeting, see Busnel 1963. In a private letter in September 1960, bioacoustician William Fish described the ICBA as defunct, as he had heard nothing further about the organization's activities since 1956 (CUL, Kellogg Papers, 3:32). Indeed, the ICBA was moribund by the early 1960s and was succeeded by the International Bioacoustics Council (IBAC), established at the Danish Natural History Museum in Aarhus in 1969. It had grown out of a small initiative started by French recordists Jean-Claude and Helen Roche of the ECHO Institute (Enregistrements et Etudes des Chants et Cris d'Oiseaux) two years earlier. Initially a group of like-minded European wildlife recordists, the council grew into an academic platform for bioacousticians in the decades that followed. Although the council still exists, its organization was transferred to the British Library Wildlife Sound Archive in London in 1996 (interview with Cheryl Tipp by the author on April 23, 2009, ibac.info/history.html).

4. These included pioneering bioacousticians such as Donald Borror at Ohio State University, William Gunn in Toronto, and Günter Tembrock at the Humboldt University Berlin.

5. In Denmark, Danish zoologists established a bioacoustics laboratory along with a library of sound recordings at the Natural History Museum of the University of Aarhus starting in 1961 (see Bondesen 1969). The British Institute of Recorded Sound was established in 1955 (see Fisher 1957). In the USSR, the Soviet Academy of Sciences established a Library of Wildlife Sounds in 1973 at its Institute of Biophysics in Pushchino-na-Oke, which expanded on the basis of a personal collection of recordings made by Boris Nikolayevich Veprintsev, a Russian biophysicist who specialized in cryogenics and nerve cells. From 1957, Veprintsev had assembled thousands of recordings documenting the Soviet avifauna and other animals, initially on a homemade magnetic tape recorder, starting while still a student at Moscow University. Two university collections preceded this state-funded project. The Moscow University Library of Animal Voices was founded in 1966–1969 in the Department

of Vertebrate Zoology at the Lomonosov Moscow State University, in collaboration with Veprintsev, to include all animal classes (Nikolsky 1973, 1975). A second library of sound recordings was established in 1972 at Leningrad University, building on a private collection of ornithological recordings collected (for both research and artistic purposes) since 1959 by the leading Russian ornithologist A. S. Malchevsky (Veprintsev 1980). Veprintsev himself published twenty-nine long-playing records. Together with the appearance of battery-operated tape recorders on the market, these records led to rising interest in wildlife sound recording among fellow zoologists, such as the Moscow ornithologist Rudolph Naumov, the zoologist Yuri Shivaev, and the Leningrad zoologist Irene Neufeldt, whose recordings Veprintsev included on some of his records. For a comprehensive overview of institutional wildlife sound archives and private collections known to the BBC and the British Library in the 1970s, see Boswall and Kettle 1979. These institutions were not the first attempts to gather natural sounds for scientific purposes. In Berlin, Wilhelm Doegen's original plans for the Lautarchiv, initiated in 1920, included a department dedicated to animal sounds.

6. For a characterization of postwar sound hunting as a new pastime enabled by the magnetic tape recording, see Bijsterveld 2004 and Morton 2000.

7. Key examples are Aronova, Baker, and Oreskes 2010; Baker and Millerand 2010; Edwards et al. 2011; Hilgartner 1995; Hine 2006; Kohler 2006; Leonelli 2011; Strasser 2011; Wouters and Schröder 2003; Zimmerman 2008.

8. For example, "boundary objects" (Star and Griesemer 1989), "trading zones" (Galison 1997; Gorman 2010; Marie 2008), and "border zones" (Mauz and Granjou 2013).

9. Including such titles as *Songs of Wild Birds* (1934), *Leach's Petrel* (1937a), *Wild Birds and Their Songs* (1937c), and *Birds of the North Woods* (1938). The latter two were published by the American Foundation for the Blind in New York. For a complete overview, see Boswall and Couzens 1982.

10. Published recordings in this period include Allen and Kellogg 1940, 1954, 1961, 1964; Kellogg and Allen 1942, 1950, 1951, 1952, 1958, 1959, 1960; Kellogg 1962c, 1965, 1969; Kellogg and Fassett 1953; Fassett 1957; Mason, Kellogg, and Allen 1954.

11. Examples are the Swedish field recordist Sture Palmér, who began recording birdsongs in the 1930s for Swedish Radio; Danish zoologist Carl Weismann, who recorded for Danish radio since the 1930s; and Ed Boyes, who was chief engineer at a Detroit radio station and past president of the Audubon Society.

12. Interview with John F. Burton by Christopher Parsons, May 3, 2001; interview with Desmond Hawkins by Christopher Parsons, October 9, 1998. Transcripts are available at WildFilmHistory.org.

13. For the BBC's attention to science as part of its public service commitment, see LaFollette 2008, 90.

14. "Extract from Max Nicholson's Opening Address at the Joint Meeting of the Linnean Society, the British Trust for Ornithology, and the British Ornithologists' Union," n.d., WAC, BBC Records, R-46-364-3/2B. Emphasis added.

15. Once again, the extent to which such liaisons could be built seemed limited. With Thorpe, the panel was also embedded in international birdsong research. When the bioacoustics community convened in Pennsylvania, for instance, Thorpe invited the BBC to associate itself with the planned International Institute of Bio-Sound, but the BBC administrators eventually felt that this would lead too far away from their initial "scientific engagements" ("Minutes of the 6th BBC Advisory Panel on Bird Recording," November 2, 1955, WAC, BBC Records, R46-719-1).

16. Simms's mandate was also limited. Although he had wanted to attend the ICBA's inaugural conference in 1956, for example, his superiors did not permit him to go. The official reason was that the discussions would be too technical and not specifically on bird sound ("Address to the XIth International Congress in Basle by Eric Simms," n.d., WAC, BBC Records, R46/366/11).

17. The actual influence of such dissemination on the public is, of course, hard to ascertain. As David Goodman (2010) and Susan Douglas (2004) have pointed out, radio listening was often a routine practice or a background to family life between stretches of more attentive listening.

18. From the 1960s onward, the BBC also actively solicited promising materials for its collection not only from broadcasting organizations and commercial publishers, but also from amateur recordists. Producers at the new Natural History Unit, for instance, purchased various recordings from British recordists such as John Kirby and E. D. H. Johnson.

19. Among others: Peter Paul Kellogg to James Chapin, September 21, 1951, CUL, Kellogg Papers, 2:22.

20. See, for example, Peter Paul Kellogg to L. Irby Davis, October 15, 1954, CUL, Kellogg Papers, 3:4, and Peter Paul Kellogg to Walter Buchen, August 7, 1951, CUL, Kellogg Papers, 2:10.

21. William R. Fish to Peter Paul Kellogg, September 19, 1951, CUL, Kellogg Papers, 3:34.

22. Interview with Günter Tembrock by the author on February 10, 2009; see also Thielcke 1961.

23. William R. Fish to Peter Paul Kellogg, September 19, 1951, CUL, Kellogg Papers, 3:34.

24. David D. Keck to Peter Paul Kellogg, April 15, 1960, CUL, Kellogg Papers, 6:16.

25. For a comprehensive history of prewar struggles for "professional jurisdiction" in ornithology, particularly between systematists and field ornithologists, see Barrow 1998.

26. Peter Paul Kellogg to Eric Simms, January 22, 1957, WAC, BBC Records, R46-365-2.

27. Peter Paul Kellogg to Ernst Booth, May 10, 1960, CUL, Kellogg Papers, 1:20.

28. Memo "Laboratory of Ornithology, Cornell University, Cumulative Financial Report for the Period Ending January 31, 1960", n.d., CUL, Kellogg Papers, 2:18.

29. Lancaster and Johnson 1976, quoted in Little 2003 194–198. The balance between these sources shifted with the expansion of the laboratory. By 1975, escalating operating costs had reduced the share of royalties in the institute's overall budget to around 10 percent, making up $25,750 alongside an annual grant from the university, membership fees, bookstore sales, and home study courses.

30. Paul Schwartz to Peter Paul Kellogg, October 14, 1963, CUL, Kellogg Papers, 7:22.

31. Memo "Pirating of Bird Sound Recordings Belonging to Cornell," by Peter Paul Kellogg, February 1958, CUL, Kellogg Papers, 2:19.

32. Peter Paul Kellogg to Jerry Stillwell, August 28, 1952, CUL, Stillwell Papers, 1:4; L. Irby Davis to Peter Paul Kellogg, January 22, 1959, CUL, Kellogg Papers, 3:3; Peter Paul Kellogg to Myles E. North, November 18, 1955, CUL, Kellogg Papers, 6:22.

33. Desmond Hawkins to Eric Simms, March 14, 1955, WAC, BBC Records, R46-719-1.

34. Peter Paul Kellogg to Eric Simms, October 21, 1954, CUL, Kellogg Papers, 2:8.

35. Peter Paul Kellogg to Eric Simms, January 26, 1955, CUL, Kellogg Papers, 2:8.

36. Peter Paul Kellogg to Timothy Eckersley, June 13, 1956, WAC, BBC Records, R46-365-2.

37. Timothy Eckersley to Peter Paul Kellogg, July 3, 1956, WAC, BBC Records, R46-365-2.

38. Peter Paul Kellogg to L. Irby Davis, November 18, 1957, CUL, Kellogg Papers, 3:4.

39. Peter Paul Kellogg to Jerry Stillwell, January 19, 1952, CUL, Stillwell Papers, 1:4.

40. Peter Paul Kellogg to William Fish, November 20, 1952, CUL, Kellogg Papers, 3:34.

41. See, for instance, William Sladen to Peter Paul Kellogg, April 28, 1960, CUL, Kellogg Papers, 7:14; Peter Paul Kellogg to Myles E. North, December 1951, CUL, Kellogg Papers, 6:24.

42. Peter Paul Kellogg to Jerry Stillwell, May 29, 1952, CUL, Stillwell Papers, 1:2.

43. See Davis 1958b; Fish 1954; North and McChesney 1964; Reynard and Kellogg 1969; Stillwell and Stillwell 1961; Schwartz 1963.

44. See, for instance, Peter Paul Kellogg to Pershing B. Hofflund, July 26, 1961, CUL, Kellogg Papers, 4:1; Peter Paul Kellogg to Jeffrey Boswall, July 9, 1961, CUL, Kellogg Papers, 2:1; Peter Paul Kellogg to Donald McChesney, October 22, 1958, CUL, Kellogg Papers, 5:46.

45. Myles E. North to Peter Paul Kellogg, July 8, 1963, CUL, Kellogg Papers, 6:19.

46. Donald S. McChesney to Peter Paul Kellogg, October 23, 1958, CUL, Kellogg Papers, 5:46.

47. Donald S. McChesney to Peter Paul Kellogg, October 22, 1958, CUL, Kellogg Papers, 5:46.

48. Peter Paul Kellogg to L. Irby Davis, March 20, 1952, and February 24, 1954, CUL, Kellogg Papers, 3:4.

49. Myles E. North to Peter Paul Kellogg, September 24, 1952, CUL, Kellogg Papers, 6:24.

50. Robert C. Stein to John W. Hardy, November 27, 1962, CUL, Kellogg Papers, 1:9. See also Peter Paul Kellogg to Ernest Booth, May 10, 1960, CUL, Kellogg Papers, 1:20; Peter Paul Kellogg to Edgar Queeny, December 9, 1950, CUL, Kellogg Papers, 2:22.

51. Alfred J. Evrier to Peter Paul Kellogg, n.d., CUL, Kellogg Papers, 1:9. See also Bruyninckx 2012.

52. See, for instance, Peter Paul Kellogg to Jerry Stillwell, July 17, 1952, CUL, Stillwell Papers, 1:44.

53. William Fish to Jerry Stillwell, August 19, 1954, CUL, Stillwell Papers, 1:28.

54. Peter Paul Kellogg to William H. Thorpe, November 21, 1955, CUL, Kellogg Papers, 8:8.

55. Myles E. North to Peter Paul Kellogg, February 12, 1963, CUL, Kellogg Papers, 6:20.

56. Peter Paul Kellogg to Jerry Stillwell, October 3, 1952, CUL, Stillwell Papers, 1:5. See also, for example, Peter Paul Kellogg to L. Irby Davis, March 25, 1955, CUL, Kellogg Papers, 3:4.

57. See, for instance, Peter Paul Kellogg to L. Irby Davis, August 27, 1956, CUL, Kellogg Papers, 3:4; interview by the author with Greg Budney (CLNS curator), September 10, 2010.

58. Interview by the author with Randolph Scott Little (former CLNS), October 10, 2010, and personal communication, April 12, 2011.

59. Ed Boyes to Peter Paul Kellogg, November 12, 1951, CUL, Kellogg Papers, 1:37.

60. For example, Peter Paul Kellogg to E. M. Barraud, June 27, 1960, CUL, Kellogg Papers, 1:20.

61. For instance, Peter Paul Kellogg to James Chapin, January 2, 1951, CUL, Kellogg Papers, 2:22.

62. Interview by the author with Randolph Scott Little, October 10, 2010.

63. Peter Paul Kellogg to Walter Buchen, August 7, 1951, CUL, Kellogg Papers, 2:10.

64. See also Biagioli and Galison 2003.

65. Peter Paul Kellogg to Jerry Stillwell, April 7, 1951, CUL, Stillwell Papers, 1:2.

66. Peter Paul Kellogg to Donald S. McChesney, November 17, 1961, CUL, Kellogg Papers, 5:43.

5   Patterned Sound

1. The microscope proved a popular trope. Reviewing a hundred years of birdsong research, biologist Myron Baker (2001, 3), for instance, also likened the "new horizons in bird song studies" opened up by the sound spectrograph to the "new world of organisms revealed by the first microscope." The implication of these analogies is that like the microscope, sound spectrography ultimately gave biologists access to more sophisticated resolution in their analysis.

2. I draw here in part on an insightful and beautifully illustrated (and audified) overview of early attempts at inscribing sound; see Feaster 2012.

3. The Kay Electric Company's "Sona-Graph" is not to be confused with the "sonograph," which was developed independently by a Swiss engineer, Jean Dreyfus-Graf. Like the sound spectrograph, the sonograph was a frequency analyzer devised to aid speech recognition, but its approach was different. The sonograph produced a cursive, seemingly stenographic representation of speech, which the inventor suggested could be converted to a phonetic typewriter so that each spoken word could ultimately be identified automatically, though these further developments were never successful. Although researchers at Bell Labs had considered automatic speech recognition based on the detection of spectral patterns for phrases, this had focused on subparts (phonemes, syllables, or parts of syllables) (Ohala 2014).

4. Ralph K. Potter to Arthur A. Allen, June 29, 1948, CUL, Kellogg Papers, 1:25. This collection of amphibian recordings was the felicitous result of a contracted research project during the war. See chapter 3, note 9.

5. Although methodological debate had declined since the late 1920s, the problem of sound transcription remained on the agenda. By 1950, for instance, Myles North had published a short paper arguing for the continued relevance of transcription for the observer and suggesting a stripped-down musical notation for use in identification and field study. Acknowledging the recent advances in birdsong recording and its physical analysis, particularly at Cornell, he suggested the mutual relevance of comparing song analysis by "accurate instruments" with "aural transcripts" (North 1950, 101).

6. For more on developments in the Cambridge University Sub-Department of Animal Behaviour, see Radick 2007; Marler 1985; Hinde 1985.

7. Marler had been appointed there to set up a project investigating song learning in juncos and white-crowned sparrows, together with graduate students Miwako Tamura and Masukazu Konishi (Konishi 1964; Marler 1959, 1960; Marler and Isaac 1960a, 1960b; Marler and Tamura 1962).

8. Donald J. Borror to Peter Paul Kellogg, October 12, 1961, CUL, Kellogg Papers, 1:22.

9. Peter Paul Kellogg to William R. Fish, July 31, 1962, CUL, Kellogg Papers, 3:32.

10. Peter Paul Kellogg to Crawford H. Greenewalt, October 11, 1962, CUL, Kellogg Papers, 3:39.

11. Similar experiments, among others by Friedrich Vilbig, were reported by Dennis Gabor (1952) in his "Lectures on Communication Theory."

12. See Grimes 1966, Kroodsma and Miller 1980, Thompson 1970, or Thorpe 1961b.

13. Borror published a review of Thorpe's Bird-Song a few months after it appeared in the Wilson Bulletin, noting: "The book is profusely illustrated with sound spectrographs. These graphs are produced in black and white, and are often somewhat diagrammatic rather than exact reproductions of the original graphs. The result is that many low amplitude sounds are lost, and amplitude variations within the song are not indicated—features which in my opinion detract from the value of the graphs. The time scale in a few graphs is too small to show many of the finer details. The author occasionally misinterprets his graphs, e.g. in Fig. 51, in indicating the frequencies of the energy peaks" (Borror 1962, 107).

14. Peter Paul Kellogg to Myles E. North, January 30, 1962, CUL, Kellogg Papers, 6:19.

15. Peter Paul Kellogg to Gustav A. Swanson, December 11, 1961, CUL, Kellogg Papers, 8:1.

16. Peter Paul Kellogg to Donald J. Borror, October 26, 1961, CUL, Kellogg Papers, 1:22.

17. Peter Paul Kellogg to L. Irby Davis, April 20, 1962, CUL, Kellogg Papers, 3:3.

18. Richard E. Barthelemy to Peter Paul Kellogg, August 3, 1961, CUL, Kellogg Papers, 1:22.

19. L. Irby Davis to Peter Paul Kellogg, February 5, 1962, CUL, Kellogg Papers, 3:3.

20. L. Irby Davis to Peter Paul Kellogg, February 15, 1961, CUL, Kellogg Papers, 3:3; Peter Paul Kellogg to Crawford H. Greenewalt, February 13, 1961, CUL, Kellogg Papers, 3:39; Peter Paul Kellogg to L. Irby Davis, December 2, 1960, CUL, Kellogg Papers, 3:3, and on.

21. This debate has an interesting parallel with present-day concerns within the scientific community regarding routine alterations to digital images, which have led high-profile science journals to introduce guidelines for authors regarding image manipulation and forensic procedures to detect misuse. The debate surrounding spectrographic adjustments shows that these controversies are not unique to digital image processing. For a discussion of how digital image processing recalibrates existing tensions and ambiguities in image production and processing, see Frow 2012.

22. Letter CaUL, Thorpe Papers, Ms.Add. 8784 W.H.T. Set 8ii Folder II.

23. This point is explored in detail by Gregory Radick (2007, 270–295).

24. Consider, for instance, the influential volume that Lanyon and Tavolga (1960) edited on *Animal Sounds and Communication*.

25. The manufacturers of commonly used spectrographs, such as Kay Electronics, later added a logarithmic display module to their commercial sound spectrographs, but this setting was rarely used in practice.

26. See, for instance, Glue 1978, 125.

27. A similar debate has animated ethnomusicology, especially in the second half of the twentieth century. Following up on Milton Metfessel's mechanical transcriptions, the introduction of melographic devices in ethnomusicology was heralded by proponents such as Charles Seeger as a new era for the field. But as Bruno Nettl (2015, 82–86) has pointed out, the devices did not supplant human efforts at transcription in part because musicologists were unable to derive from them directly a sense of musical sound. In the 1960s and 1970s, the prevailing view was that trained ethnomusicologists remained important, because as Jairazbhoy (1977) points out, a large gap exists between what an automatic transcriber "hears" and what an experienced listener in a particular idiom hears.

6 Conclusion

1. It continues to be a topic of debate, however, whether the increase in minimum frequency in urban birds' vocalizations is a direct adaptation to their surroundings or an epiphenomenon of an increased amplitude—a well-known side effect of the Lombard effect. See Slabbekoorn, Xiao-Jing, and Halfwerk 2012 as well as Nemeth and Brumm 2010.

2. This development seems to have emerged separately from acoustic investigations of noise control by means of vegetation. Once the forest had been defined as an acoustic environment, acoustic engineers also began to apply their expertise for various nonmilitary purposes. In the 1960s and 1970s, for instance, urban planners experimented with trees and vegetation as sound attenuators in West Berlin (Jasper 2015, 82–91).

References

Source Collections and Their Abbreviations

Allen Papers (CUL): Arthur A. Allen Papers, #21-18-1255, Division of Rare and Manuscript Collections, Carl A. Kroch Library, Cornell University Library.

BBC NHRU (WAC): British Broadcasting Corporation, Recording General and Natural History Recording Unit files, Record Units R46, R57, S26, British Broadcasting Corporation, Written Archives Center.

BLOWS Records (BL): British Library of Wildlife Sounds, Records, 1969– (ongoing).

Brand Papers (CUL): Albert R. Brand Papers, #21-18-899, Division of Rare and Manuscript Collections, Carl A. Kroch Library, Cornell University Library.

Kellogg Papers (CUL): Peter P. Kellogg Papers, #21-18-893, Division of Rare and Manuscript Collections, Carl A. Kroch Library, Cornell University Library.

MLNS Papers (CLO): Cornell Laboratory of Ornithology, Macaulay Library of Natural Sounds, Records, 1931– (ongoing).

Stillwell Papers (CUL): Jerry and Norma Stillwell Papers, #2621, Division of Rare and Manuscript Collections, Carl A. Kroch Library, Cornell University Library.

Tanner Papers (CUL): James Taylor Tanner Papers, #2665, Division of Rare and Manuscript Collections, Carl A. Kroch Library, Cornell University Library.

Thorpe Papers (CaUL): William Homan Thorpe Papers, GB 012 Ms.Add 8784, Manuscript Collections, Cambridge University Library.

Published Sources

Abbott, A. 1988. *The System of Professions: An Essay on the Division of Expert Labor.* Chicago: University of Chicago Press.

Abbott, C. C. 1896. *Bird-Land Echoes.* Philadelphia: Lippincott.

Ainley, M. G. 1979. "The Contribution of the Amateur to North American Ornithology: A Historical Perspective." *Living Bird* 18 (1): 161–177.

Albert R. Brand Foundation. 1948. *Voices of the Night: The Calls of 26 Frogs and Toads Found in Eastern North America.* Ithaca, NY: Comstock Publishing Company. BL-10743. Four 78 rpm discs.

Alexander, H. G. 1943. "Report on the Bird-Song Inquiry." *British Birds* 364:65–72.

Allen, A. A. 1933. "Ornithological Education in America." In *Fifty Years' Progress of American Ornithology, 1883–1933*, ed. Frank M. Chapman and T. S. Palmer, 215–229. Lancaster, PA: American Ornithologists' Union.

Allen, A. A., and P. P. Kellogg. 1940. *Do You Know the Birds?* Albany: New York State Education Department. Two 78 rpm discs.

Allen, A. A., and P. P. Kellogg. 1954. *Songbirds of America in Color, Sound and Story.* New York: Book Records Inc. Book and 33⅓ rpm disc.

Allen, A. A., and P. P. Kellogg. 1961. *Bird Songs in Your Garden.* New York: Cornell University Laboratory of Ornithology. Book and 33⅓ rpm disc.

Allen, A. A., and P. P. Kellogg. 1964. *Bird Songs of Garden, Woodland and Meadow.* Washington, DC: National Geographic Society. Book and six 33⅓ rpm discs.

Allen, D. E. 1976. *The Naturalist in Britain.* Princeton, NJ: Princeton University Press.

Allen, E. G. 1951. "The History of American Ornithology before Audubon." *Transactions of the American Philosophical Society* 41 (3): 387–591.

Allen, F. H. 1919. "The Evolution of Bird-Song" *Auk* 36 (4): 528–536.

Alpers, S. 1998. "The Studio, the Laboratory, and the Vexations of Art." In *Picturing Science, Producing Art*, ed. C. A. Jones and P. Galison, 401–419. New York: Routledge.

Alström, P., and R. Ranft. 2003. "The Use of Sounds in Avian Systematics and the Importance of Bird Sound Archives." *Bulletin of the British Ornithologists' Club* 123A:114–135.

Altman, R. 1992. "Sound Space." In *Sound Theory/Sound Practice*, ed. R. Altman, 46–64. New York: Routledge.

Amann, K., and K. Knorr Cetina. 1988. "The Fixation of (Visual) Evidence." *Human Studies* 11 (2/3): 133–169.

Ames, E. 2003. "The Sound of Evolution." *Modernism/Modernity* 10 (2): 297–325.

Andrew, R. J. 1957. "A Comparative Study of the Calls of Emberiza SPP (Buntings)." *Ibis* 99 (1): 27–42.

Ansley, H., and S. Ansley. 1961. *The Birds World of Song.* New York: Folkways Records/American Museum of Natural History. 78 rpm disc.

Armitage, K. C. 2009. *The Nature Study Movement: The Forgotten Populizer of America's Conservation Ethic*. Lawrence: University of Kansas Press.

Aronova, E., K. S. Baker, and N. Oreskes. 2010. "Big Science and Big Data in Biology: From the International Geophysical Year through the International Biological Program to the Long Term Ecological Research (LTER) Network, 1957–Present."*Historical Studies in the Natural Sciences* 40 (2): 183–224.

Axtell, H. H. 1938. "The Song of Kirtland's Warbler." *Auk* 55 (3): 481–491.

Bailey, C. E. G. 1950. "Towards an Orthography of Bird Song." *Ibis* 92 (1): 115–122.

Baker, K. S., and F. Millerand. 2010. "Infrastructuring Ecology: Challenges in Achieving Data Sharing." In *Collaboration in the New Life Sciences*, ed. J. N. Parker, N. Vermeulen, and B. Penders, 111–138. Farnham: Ashgate.

Baker, M. C. 2001. "Bird Song Research: The Past 100 Years." *Bird Behaviour* 14 (1): 3–50.

Baptista, L. F., and S. L. L. Gaunt. 1997. "Bioacoustics as a Tool in Conservation Studies." In *Behavioral Approaches to Conservation in the Wild*, ed. J. R. Clemmons and R. Buchholz, 212–242. Cambridge: Cambridge University Press.

Barrow, M. V. 1998. *A Passion for Birds: American Ornithology after Audubon*. Princeton, NJ: Princeton University Press.

Barrow, M. V. 2000. "The Specimen Dealer: Entrepreneurial Natural History in America's Gilded Age." *Journal of the History of Biology* 33 (3): 493–534.

Barrow, M. V. 2011. "On the Trail of the Ivory-Bill: Field Science, Local Knowledge, and the Struggle to Save Endangered Species." In *Knowing Global Environments: New Historical Perspectives on the Field Sciences*, ed. J. Vetter, 135–161. New Brunswick, NJ: Rutgers University Press.

Batchellor, D. 1913. "Musical Kindergarten Method." In *Young Folk's Handbook*, ed. American Institute of Child Life, 45. Philadelphia: American Institute of Child Life.

Battalio, J. T. 1998. *The Rhetoric of Science in the Evolution of American Ornithological Discourse*. Vol. 8. London: Ablex.

Bell, S., M. Marzano, J. Cent, H. Kobierska, D. Podjed, D. Vandzinskaite, H. Reinert, A. Armaitiene, M. Grodzińska-Jurczak, and R. Muršič. 2008. "What Counts? Volunteers and Their Organisations in the Recording and Monitoring of Biodiversity." *Biodiversity and Conservation* 17:3443–3454.

Bevis, J. 2010. *aaaaw to zzzzzd: The Words of Birds*. Cambridge, MA: MIT Press.

Beyer, G. G. 1900. "The Ivory-Billed Woodpecker in Louisiana." *Auk* 17 (2): 97–99.

Beyer, R. T. 1999. *Sounds of Our Times: Two Hundred Years of Acoustics*. New York: Springer.

Biagioli, M. 1998. "The Instability of Authorship: Credit and Responsibility in Contemporary Biomedicine." *FASEB Journal* 12 (1): 3–16.

Biagioli, M., and P. Galison, eds. 2003. *Scientific Authorship: Credit and Intellectual Property in Science.* New York: Routledge.

Bijsterveld, K. 2001. "The Diabolical Symphony of the Mechanical Age: Technology and Symbolism of Sound in European and North American Noise Abatement Campaigns, 1900–40." *Social Studies of Science* 31 (1): 37–70.

Bijsterveld, K. 2003. "'The City of Din': Decibels, Noise, and Neighbors in The Netherlands, 1910–1980." *Osiris* 18:173–193.

Bijsterveld, K. 2004. "What Do I Do with My Tape Recorder ...?" Sound Hunting and the Sounds of Everyday Dutch Life in the 1950s and 1960s." *Historical Journal of Film, Radio and Television* 24 (4): 613–634.

Bijsterveld, K. 2008. *Mechanical Sound: Technology, Culture, and Public Problems of Noise in the Twentieth Century.* Cambridge, MA: MIT Press.

Bircham, P. 2007. *A History of Ornithology.* London: Collins.

Birkhead, T. R. 2003. *A Brand-New Bird: How Two Amateur Scientists Created the First Genetically Engineered Animal.* New York: Basic Books.

Birkhead, T. R. 2011. *The Wisdom of Birds: An Illustrated History of Ornithology.* London: Bloomsbury. Princeton, NJ: Princeton University Press.

Birkhead, T. R., and S. van Balen. 2008. "Bird-Keeping and the Development of Ornithological Science. *Archives of Natural History* 35 (2): 281–305.

Birkhead, T. R., J. Wimpenny, and B. Montgomerie. 2014. *Ten Thousand Birds: Ornithology since Darwin.* Princeton, NJ: Princeton University Press.

Bondesen, P. 1969. "A Bio-acoustic Laboratory in Denmark." *Recorded Sound* 34:449–450.

de Bont, R. 2009. "Poetry and Precision: Johannes Thienemann, the Bird Observatory in Rossitten and Civic Ornithology, 1900–1930." *Journal of the History of Biology* 44 (2): 171–203.

de Bont, R. 2015. *Stations in the Field: A History of Place-Based Animal Research.* Chicago: University of Chicago Press.

Borror, D. J. 1956. "Variation in Carolina Wren Songs." *Auk* 73 (2): 211–229.

Borror, D. J. 1959. "Variation in the Songs of the Rufous-Sided Towhee." *Wilson Bulletin* 71 (1): 54–72.

Borror, D. J. 1961. "Intraspecific Variation in Passerine Bird Songs." *Wilson Bulletin* 73 (1): 57–78.

Borror, D. J. 1962. "Review of *Bird Song: The Biology of Vocal Communication and Expression in Birds*." *Wilson Bulletin* 74 (1): 107–108.

Borror, D. J., and C. R. Reese. 1953. "The Analysis of Bird Songs by Means of a Vibralyzer." *Wilson Bulletin* 65 (4): 271–303.

Boswall, J., and D. Couzens. 1982. "Fifty Years of Bird Sound Publication in North America: 1931–1981." *North American Birds* 36 (6): 924–943.

Boswall, J., and R. Kettle. 1979. "A Revised World List of Wildlife Sound.' *Recorded Sound* 74/75:70–72.

Bourdieu, P. 1986. "The Forms of Capital." In *Handbook of Theory and Research for the Sociology of Education*, ed. J. Richardson, 241–258. New York: Greenwood.

Brady, E. 1999. *A Spiral Way: How the Phonograph Changed Ethnography*. Jackson: University Press of Mississippi.

Brain, R. M. 1998. "Standards and Semiotics." In *Inscribing Science: Scientific Texts and the Materiality of Communication*, ed. T. Lenoir, 249–285. Stanford, CA: Stanford University Press.

Brain, R. M. 2015. *The Pulse of Modernism: Physiological Aesthetics in Fin-de-Siècle Europe*. Seattle: University of Washington Press.

Brand, A. R. 1932. "Recording Sounds of Wild Birds." *Auk* 49 (4): 436–449.

Brand, A. R. 1934. *Songs of Wild Birds*. New York: Thomas Nelson and Sons. Book and two 78 rpm discs.

Brand, A. R. 1935. "A Method for the Intensive Study of Bird Song." *Auk* 52 (1): 40–52.

Brand, A. R. 1936. *More Songs of Wild Birds*. New York: Thomas Nelson and Sons. Book and three 78 rpm discs.

Brand, A. R. 1937a. *Leach's Petrel*. New York: Albert R. Brand. 78 rpm disc.

Brand, A. R. 1937b. "Why Bird Song Can Not Be Described Adequately." *Wilson Bulletin* 49 (1): 11–14.

Brand, A. R. 1937c. *Wild Birds and Their Songs*. New York: American Foundation for the Blind. Two 78 rpm discs.

Brand, A. R. 1938. *Birds of the North Woods*. New York: American Foundation for the Blind. Two 33⅓ rpm discs.

Brand, A. R., and H. Axtell. 1938. "Song of the Chipping Sparrow." *Auk* 55 (1): 125–126.

Brauner, J. 1979. "In Memoriam: Aretas Andrew Saunders." *Auk* 96 (1): 172–178.

Breittruck, J. 2012. "Pet Birds: Cages and Practices of Domestication in Eighteenth-Century Paris." *InterDisciplines* 1:6–24.

Bruun, B., H. Delin, and L. Svensson. 1992. *Birds of Britain and Europe*. London: Hamlyn.

Bruyninckx, J. 2012. "Sound Sterile: Making Scientific Field Recordings in Ornithology." In *The Oxford Handbook of Sound Studies*, ed. T. Pinch and K. Bijsterveld, 127–150. Oxford: Oxford University Press.

Burkhardt, R. W., Jr. 1992. "Huxley and the Rise of Ethology." In *Julian Huxley: Biologist and Statesman of Science*, ed. K. C. Waters and A. V. Helden, 127–149. Houston, TX: Rice University Press.

Burkhardt, R. W., Jr. 1999. "Ethology, Natural History, the Life Sciences, and the Problem of Place." *Journal of the History of Biology* 32 (3): 489–508.

Burkhardt, R. W., Jr. 2005. *Patterns of Behavior: Konrad Lorenz, Niko Tinbergen, and the Founding of Ethology*. Chicago: University of Chicago Press.

Burnett, G. D. 2012. *The Sounding of the Whale: Science and Cetaceans in the Twentieth Century*. Chicago: University of Chicago Press.

Burri, R. V., and J. Dumit. 2008. "Social Studies of Scientific Imaging and Visualization." In *The Handbook of Science and Technology Studies*, 3rd ed., ed. E. J. Hackett, O. Amsterdamska, M. Lynch, and J. Wajcman, 297–317. Cambridge, MA: MIT Press.

Burroughs, J. 1903. "Real and Sham Natural History." *Atlantic Monthly* 91 (March): 298.

Burton, J. F. 1967. "The Scientific Use of BBC Wildlife Sound Recordings. *Ibis* 109 (1): 126.

Busnel, R.-G., ed. 1963. *Acoustic Behavior of Animals*. Amsterdam: Elsevier.

Butcher, G. S., and K. McGowan. 1995. "History of Ornithology at Cornell University." In *Contributions to the History of North American Ornithology*, ed. William E. Davis and Jerome A. Jackson, 223–245. Cambridge, MA: Nuttall Ornithological Club.

Cambrosio, A., D. Jacobi, and P. Keating. 1993. "Ehrlich's 'Beautiful Pictures" and the Controversial Beginnings of Immunological Imagery." *Isis* 84 (4): 662–699.

Canning, C. 2005. *The Most American Thing in America: Circuit Chautauqua as Performance*. Iowa City: University of Iowa Press.

Carson, R. 1962. *Silent Spring*. London: Hamilton.

Catchpole, C. K. 1973. "The Functions of Advertising Song in the Sedge Warbler (*Acrocephalus schoenobaenus*) and the Reed Warbler (*A. scirpaceus*)." *Behaviour* 46 (3/4): 300–320.

Catchpole, C. K., and P. J. B. Slater. 1995. *Bird Song: Biological Themes and Variations*. Cambridge: Cambridge University Press.

Cheney, S. P. 1892. *Wood Notes Wild: Notations of Bird Music*. Boston: Lee and Shepard.

Clark, X. 1879. "Animal Music, Its Nature and Origin." *American Naturalist* 13 (4): 209–223.

Clarke, A., and J. Fujimura, eds. 1992. *The Right Tools for the Job: At Work in Twentieth-Century Life Sciences*. Princeton, NJ: Princeton University Press.

Classen, C. 1993. *Worlds of Sense: Exploring the Senses in History and Across Cultures*. London: Routledge.

Coates, P. A. 2005. "The Strange Stillness of the Past: Toward an Environmental History of Sound and Noise." *Environmental History* 10 (4): 636–665.

Coffin, L. V. B. 1928. "Individuality in Bird Song." *Wilson Bulletin* 40 (2): 95–99.

Coleman, J. S. 1988. "Social Capital in the Creation of Human Capital." *American Journal of Sociology* 94:S95–S120.

Collias, N. E., and M. Joos. 1953. "The Spectrographic Analysis of Sound Signals of the Domestic Fowl." *Behaviour* 5 (3): 175–188.

Collins, H. M. 2001. "Tacit Knowledge, Trust, and the Q of Sapphire." *Social Studies of Science* 31 (1): 71–86.

Comstock, A. B. [1911] 1939. *Handbook of Nature Study*. Ithaca, NY: Comstock Publishing Associates.

Cooper, F. S., A. M. Liberman, and J. M. Borst. 1951. "The Interconversion of Audible and Visible Patterns as a Basis for Research in the Perception of Speech." *Proceedings of the National Academy of Sciences of the United States of America* 37 (5): 318–325.

Coopmans, C., J. Vertesi, M. Lynch, and S. Woolgar, eds. 2014. *Representation in Scientific Practice Revisited*. Cambridge, MA: MIT Press.

Copeland, P., and J. Boswall. 1983. "A Discography of Human Imitation of Bird Sound." *Recorded Sound* 83:73–100.

Copeland, P., J. Boswall, and L. Petts. 1988. *Birdsongs on Old Records: A Coarsegroove Discography of Palearctic Region Bird Sound 1910–1958*. London: British Library National Sound Archive Wildlife Section.

Corbin, A. 1998. *Village Bells: Sound and Meaning in the Nineteenth-Century French Countryside*. Trans. M. Thom. New York: Columbia University Press.

Craft, R., and Igor Stravinsky. 1952. *Conversations with Igor Stravinsky*. Garden City, NY: Doubleday.

Cramp, S., and K. E. L. Simmons, eds. 1977. *Handbook of the Birds of Europe, the Middle East and North Africa: The Birds of the Western Palearctic*, Vol. 1: *Ostrich to Ducks*. Oxford: Oxford University Press.

Crist, E. 1996. "Naturalists' Portrayals of Animal Life: Engaging the Verstehen Approach." *Social Studies of Science* 24:799–838.

Crist, E. 1999. *Images of Animals: Anthropomorphism and Animal Mind*. Philadelphia: Temple University Press.

Dalton, R. 2005. "A Wing and a Prayer." *Nature* 2005 (September 8): 188–190.

Darwin, C. 1871. *The Descent of Man and Selection in Relation to Sex*. London: John Murray.

Daston, L. 1995. "The Moral Economy of Science." *Osiris* 10:3–24.

Daston, L., and P. Galison. 2007. *Objectivity*. New York: Zone Books.

Daston, L., and E. Lunbeck, eds. 2011. *Histories of Scientific Observation*. Chicago: University of Chicago Press.

Daston, L., and H. O. Sibum. 2003. "Introduction: Scientific Personae and Their Histories." *Science in Context* 16 (1/2): 1–8.

Davies, G. 2000. "Science, Observation and Entertainment: Competing Visions of Postwar British Natural History Television, 1946–1967."*Cultural Geographies* 7 (4): 432–460.

Davis, L. I. 1958a. "Acoustic Evidence of Relationship in North American Crows." *Wilson Bulletin* 70 (2): 151–167.

Davis, L. I. 1958b. *Mexican Bird Songs*. Ithaca, NY: Cornell University Laboratory of Ornithology. 33⅓ rpm disc.

Davis, L. I. 1964. "Biological Acoustics and the Use of the Sound Spectrograph." *Southwestern Naturalist* 9 (3): 118–145.

Davis, S. 2011. "Militarised Natural History: Tales of the Avocet's Return to Postwar Britain." *Studies in History and Philosophy of Biological and Biomedical Sciences* 42 (2): 226–232.

Davis, W. E., Jr. 1994. *Dean of the Birdwatchers: A Biography of Ludlow Griscom*. Washington, DC: Smithsonian Institution Press.

Davis, W. E., Jr., and J. A. Jackson, eds. 1995. *Contributions to the History of North American Ornithology*. Cambridge, MA: Nuttal Ornithological Club.

Day, T. 2002. *A Century of Recorded Music: Listening to Musical History*. New Haven, CT: Yale University Press.

Dingle, C. P., and N. Simeone. 2007. *Olivier Messiaen: Music, Art, and Literature*. Aldershot: Ashgate.

Dixon, C. 1897. *Our Favourite Song Birds: Their Habits, Music, and Characteristics*. London: Lawrence and Bullen.

Dixon, R. 1917. *The Human Side of Birds*. New York: Frederick A. Stokes.

Dobson, C. W., and R. E. Lemon. 1977. "Bird Song as Music." *Journal of the Acoustical Society of America* 61 (3): 888–890.

Doughty, R. W. 1975. *Feather Fashions and Bird Preservation: A Study in Nature Protection*. Berkeley: University of California Press.

Douglas, S. J. 1992. "Audio Outlaws: Radio and Phonograph Enthusiasts." In *Possible Dreams: Technological Enthusiasm in America*, ed. J. L. Wright, 45–49. Dearborn, MI: Henry Ford Museum and Greenfield Village.

Douglas, S. J. 2004. *Listening In: Radio and the American Imagination*. Minneapolis: University of Minnesota Press.

Dreher, C. 1931. "Microphone Concentrators in Picture Production." *Journal of the Society of Motion Picture Engineers* 14 (1): 23–30.

Dunlap, T. R. 2011. *In the Field, among the Feathered: A History of Birders and Their Guides*. Oxford: Oxford University Press.

Durant, J. R. 1986. "The Making of Ethology: The Association for the Study of Animal Behaviour, 1936–1986." *Animal Behaviour* 34 (6): 1601–1616.

Edwards, P. N., M. S. Mayernik, A. L. Batcheller, G. C. Bowker, and C. L. Borgman. 2011. "Science Friction: Data, Metadata, and Collaboration." *Social Studies of Science* 41 (5): 667–690.

Eley, C. 2013. Making Them Talk: Animals, Music, and Museums." *Antennae: The Journal of Nature and Visual Culture* 27 (1): 6–18.

Eley, C. 2014. "A Birdlike Act: Sound Recording, Nature Imitation, and Performance Whistling." *Velvet Light Trap* 74:4–15.

Ellis, R. 2011. "Jizz and the Joy of Pattern Recognition: Virtuosity, Discipline and the Agency of Insight in UK Naturalist Arts of Seeing." *Social Studies of Science* 41 (6): 769–790.

Ellis, R., and C. Waterton. 2004. "Environmental Citizenship in the Making: The Participation of Volunteer Naturalists in UK Biological Recording and Biodiversity Policy." *Science & Public Policy* 31 (2): 95–105.

Ellis, R., and C. Waterton. 2005. "Caught between the Cartographic and the Ethnographic Imagination: The Whereabouts of Amateurs, Professionals, and Nature in Knowing Biodiversity." *Environment and Planning D: Society and Space* 23 (5): 673–693.

Elway, T. 1932. "The Microphone, Radio's Electric Ear." *Radio News* 13 (7): 555–573.

Emlen, J. T., and M. J. DeJong. 1991. "Counting Birds: The Problem of Variable Hearing Abilities." *Journal of Field Ornithology* 63 (1): 26–31.

Erlmann, V. 2010. *Reason and Resonance: A History of Modern Aurality*. New York: Zone Books.

Etzkowitz, H. 1989. "Entrepreneurial Science in the Academy: A Case of the Transformation of Norms." *Social Problems* 36 (1): 14–29.

Evans, J. A. 2010. "Industry Collaboration, Scientific Sharing, and the Dissemination of Knowledge." *Social Studies of Science* 40 (5): 757–791.

Eyring, C. 1946. "Jungle Acoustics." *Journal of the Acoustical Society of America* 18 (2): 257–270.

Falls, J. B. 1992. "Playback: A Historical Perspective." In *Playback and Studies of Animal Communication*, ed. P. K. McGregor, 11–33. New York: Plenum Press.

Farrell, M. P. 2003. *Collaborative Circles: Friendship Dynamics and Creative Work*. Chicago: University of Chicago Press.

Fassett, J. H. 1957. *Symphony of the Birds (Musique Concrète)*. Old Greenwich, CO: Ficker Recording Service. C-1002:33. 33⅓ rpm disc.

Feaster, P. 2012. *Pictures of Sound: One-Thousand Years of Educed Audio, 980–1980*. Atlanta: Dust-to-Digital.

Fehr, J. 2000. "'Visible Speech' and Linguistic Insight." In *Shifting Boundaries of the Real: Making the Invisible Visible*, ed. H. Nowotny and M. Weiss, 31–49. Zurich: vdf Hochschulverlag AG an der ETH.

Fish, W. R. 1954. *Western Bird Songs*. Ithaca, NY: Cornell University Laboratory of Ornithology. 78 rpm disc.

Fisher, T. 1957. "The British Institute of Recorded Sound: A National Collection." *Tempo* 45:24–27.

Fitzpatrick, J. W., M. Lammertink, M. D. Luneau, T. W. Gallagher, B. R. Harrison, G. M. Sparling, K. V. Rosenberg, et al. 2005. "Ivory-Billed Woodpecker (*Campephilus principalis*) Persists in Continental North America.' *Science* 308 (5727): 1460–1462.

Fitzpatrick, J. W., M. Lammertink, M. D. Luneau, T. W. Gallagher, B. R. Harrison, G. M. Sparling, K. V. Rosenberg, et al. 2006. "Clarifications about Current Research on the Status of Ivory-Filled Woodpecker (*Campephilus principalis*) in Arkansas." *Auk* 123 (2): 587–593.

Fleck, L. 1979. *Genesis and Development of a Scientific Fact*. Trans. T. J. Trenn and F. Bradley. Chicago: University of Chicago Press.

Friedmann, H. 1964. "In Memoriam: Robert Thomas Moore." *Auk* 81 (3): 326–331.

Frings, H., M. Frings, B. Cox, and L. Peissner. 1955. "Auditory and Visual Mechanisms in Food-Finding Behavior of the Herring Gull." *Wilson Bulletin* 67 (3): 155–170.

Frow, E. 2012. "Drawing a Line: Setting Guidelines for Digital Image Processing in Scientific Journal Articles." *Social Studies of Science* 42 (3): 369–392.

Futuyama, D. J. 1986. "Reflections on Reflections: Ecology and Evolutionary Biology." *Journal of the History of Biology* 19 (2): 303–312.

Gabor, D. 1952. Lectures on communication theory. Research Laboratory of Electronics, MIT, April 3. hdl.handle.net/1721.1/4830.

Galison, P. L. 1997. *Image and Logic: A Material Culture of Microphysics*. Chicago: University of Chicago Press.

Garstang, W. 1922. *Songs of the Birds*. London: John Lane.

Gelatt, R. 1955. *The Fabulous Phonograph: From Tinfoil to High Fidelity*. Philadelphia: Lippincott.

Gieryn, T. F. 2002. "Three Truth-Spots." *Journal of the History of the Behavioral Sciences* 38 (2): 113–132.

Gieryn, T. F. 2006. "City as Truth-Spot: Laboratories and Field-Sites in Urban Studies." *Social Studies of Science* 36 (1): 5–38.

Gitelman, L. 2008. *Always Already New: Media, History and the Data of Culture*. Cambridge, MA: MIT Press.

Glue, D. 1978. "Review of *Handbook of the Birds of Europe, the Middle East and North Africa: The Birds of the Western Palearctic*, Vol. 1: *Ostrich to Ducks*." *Bird Study* 25 (2): 124–126.

Goodman, D. 2010. "Distracted listening: On Not Making Sound Choices in the 1930s." In *Sound in the Age of Mechanical Reproduction*, ed. D. Suisman and S. Strasser, 15–46. Philadelphia: University of Pennsylvania Press.

Goodwin, D. 1965. "A Comparative Study of Captive Blue Waxbills." *Ibis* 107 (3): 285–315.

Gorman, M. E. 2010. *Trading Zones and Interactional Expertise: Creating New Kinds of Collaboration*. Cambridge, MA: MIT Press.

Graber, R. R., and W. W. Cochran. 1959. "An Audio Technique for the Study of Nocturnal Migration of Birds." *Wilson Bulletin* 71 (3): 220–236.

Greene, P. D., and T. Porcello, eds. 2005. *Wired for Sound: Engineering and Technologies in Sonic Cultures*. Middletown, CT: Wesleyan University Press.

Greenewalt, C. H. 1968. *Bird Song: Acoustics and Physiology*. Washington, DC: Smithsonian Institution Press.

Grimes, L. 1966. "Antiphonal Singing and Call Notes of *Laniarius barbarus barbarus*." *Ibis* 108 (1): 122–126.

Griscom, L. 1922. "Problems of Field Identification." *Auk* 39 (1): 31–41.

Griscom, L. 1923. *Birds of the New York City Region*. New York: American Museum of Natural History.

Guida, M. 2015. "Britain's Sonic Therapy: Listening to Birdsong during and after the First World War." *Remedia*. http://remedianetwork.net/2015/06/16/britains-sonic -therapy-listening-to-birdsong-during-and-after-the-first-world-war/.

Haffer, J. 2001. "Ornithological Research Traditions in Central Europe during the 19th and 20th Centuries." *Journal of Ornithology* 142 (suppl. 1): 27–93.

Haffer, J. 2007. "The Development of Ornithology in Central Europe." *Journal of Ornithology* 148 (suppl. 1): 125–153.

Haffer, J. 2008. "The Origin of Modern Ornithology in Europe." *Archives of Natural History* 35 (1): 76–87.

Hagstrom, W. O. 1982. "Gift Giving as an Organizing Principle in Science." In *Science in Context*, ed. B. Barnes and D. Edge, 21–34. Cambridge, MA: MIT Press.

Hall-Craggs, J. 1962. "The Development of Song in the Blackbird." *Ibis* 104 (3): 277–300.

Hall-Craggs, J. 1979. "Sound Spectrographic Analysis: Suggestions for Facilitating Auditory Imagery." *Condor* 81 (2): 185–192.

Hall-Craggs, J. 1987. "Obituary: William Homan Thorpe." *Ibis* 129 (s2): 564–569.

Hardy, J. W. 1969. "A Taxonomic Revision of the New World Jays." *Condor* 71 (4): 360–375.

Hardy, J. W. 1975. "A Tape Recording of a Possible Ivory-Billed Woodpecker Call." *American Birds* 29 (3): 647–651.

Haring, K. 2007. *Ham Radio's Technical Culture*. Cambridge, MA: MIT Press.

Harris, A., and M. Van Drie. 2015. "Sharing Sound: Teaching, Learning and Researching Sonic Skills." *Sound Studies: An Interdisciplinary Journal* 1 (1): 98–117.

Harwood, M. 1987. "The Lab: From Hatching to Fledging." *Cornell University Laboratory of Ornithology Annual Report* 1986 (87): 4–13.

Heck, L. 1933. *Schrei der Steppe: Tönende Bilder aus dem Ostafrikanischen Busch*. Munich: Knorr and Hirth. Book and 78 rpm disc.

Heck, L. 1934. *Der Wald erschallt*. Munich: Knorr and Hirth. Book and 78 rpm disc.

Heinroth, O., and L. Koch. 1935. *Gefiederte Meistersänger*. Berlin: Hugo Vermühler. Book and three 78 rpm discs.

Henderson, K. 1991. "Flexible Sketches and Inflexible Data Bases: Visual Communication, Conscription Devices, and Boundary Objects in Design Engineering." *Science, Technology & Human Values* 16 (4): 448–473.

Henke, C. R., and T. F. Gieryn. 2008. "Sites of Scientific Practice: The Enduring Importance of Place." In *The Handbook of Science and Technology Studies*, 3rd ed., ed. E. J. Hackett, O. Amsterdamska, M. Lynch, and J. Wajcman, 353–377. Cambridge, MA: MIT Press.

Hilgartner, S. 1995. "Biomolecular Databases: New Communication Regimes for Biology." *Science Communication* 17 (2): 240–263.

Hilgartner, S., and S. I. Brandt-Rauf. 1994. "Data Dccess, Ownership, and Control: Toward Empirical Studies of Access Practices." *Knowledge: Creation, Diffusion, Utilization* 15 (4): 355–372.

Hill, G. E., D. J. Mennill, B. W. Rolek, T. L. Hicks, and K. A. Swiston. 2006. "Evidence Suggesting That Ivory-Billed Woodpeckers (*Campephilus principalis*) Exist in Florida." *Avian Conservation & Ecology* 1 (3): 2.

Hinde, R. A. 1985. "Ethology in Relation to Other Disciplines." In *Studying Animal Behavior: Autobiographies of the Founders*, ed. D. A. Dewsbury, 193–203. Chicago: University of Chicago Press.

Hine, C. 2006. "Databases as Scientific Instruments and Their Role in the Ordering of Scientific Work." *Social Studies of Science* 36 (2): 269–298.

Hjorth, I. 1970. "A Comment on Graphic Displays of Bird Sounds and Analyses with a New Device, the Melograph Mona." *Journal of Theoretical Biology* 26 (1): 1–10.

Hochman, B. 2010. "Hearing Lost, Hearing Found: George Washington Cable and the Phono-Ethnographic Ear." *American Literature* 82 (3): 519–551.

Hoffmann, B. 1908. *Kunst und Vogelgesang in Ihren Wechselseitigen Beziehungen vom Naturwissenschaftlich-Musikalischen Standpunkte*. Leipzig: Quelle and Meyer.

Hold, T. 1970. "The Notation of Bird-Song: A Review and a Recommendation." *Ibis* 112 (2): 151–172.

Homans, G. C. 1958. "Social Behavior as Exchange." *American Journal of Sociology* 63 (6): 597–606.

Homans, G. C. 1974. *Social Behavior: Its Elementary Forms*. New York: Harcourt, Brace, and World.

Hornbostel, E. M. von. 1911. "Musikpsychologische Bemerkungen über Vogelge-sang." *Zeitschrift der Internationalen Musikgesellschaft* 12:117–128.

Hughes, T. P. 1983. *Networks of Power: Electrification in Western Society, 1880–1930*. Baltimore: Johns Hopkins University Press.

Hui, A. 2013. *The Psychophysical Ear: Musical Experiments, Experimental Sounds, 1840–1910*. Cambridge, MA: MIT Press.

Hui, A., J. Kursell, and M. W. Jackson. 2013. *Music, Sound, and the Laboratory from 1750–1980*. Chicago: University of Chicago Press.

Hunt, F. V. 1978. *Origins in Acoustics: The Science of Sound from the Age of Newton*. New Haven, CT: Yale University Press.

Hunt, R. 1922. "Evidence of Musical 'Yaste' in the Brown Towhee." *Condor* 24 (6): 193–203.

Huxley, J. S. 1916. "Bird-Eatching and Niological Dcience: Some Observations on the Dtudy of Vourtship in Birds." *Auk* 33 (2): 142–161.

Huxley, J. S. 1949. *Birdwatching and Bird Behaviour*. London: Dobson.

Huxley, J. S., and L. Koch. 1938. *Animal Language*. London: Country Life Ltd.

Jackson, M. W. 2006. *Harmonious Triads: Physicists, Musicians, and Instrument Makers in Nineteenth-Century Germany*. Cambridge, MA: MIT Press.

Jacob, M. C. 1999. "Science Studies after Social Construction: The Turn toward the Comparative and the Global." In *Beyond the Cultural Turn: New Directions in the Study of Society and Culture*, ed. V. E. Bonnell and L. Hunt, 95–120. Berkeley: University of California Press.

Jacobs, N. J. 2016. *Birders of Africa: History of a Network*. New Haven, CT: Yale University Press.

Jairazbhoy, N. A. 1977. "The 'Objective' and Subjective View in Music Transcription." *Ethnomusicology* 21 (2): 263–273.

Jasper, S. 2015. "Cyborg Imaginations: Nature, Technology, and Urban Space in West Berlin." Doctoral dissertation, University College London.

Johnson, J. H. 1995. *Listening in Paris: A Cultural History*. Berkeley: University of California Press.

Johnson, K. 2004. "*The Ibis*: Transformations in a Twentieth Century British Natural History Journal." *Journal of the History of Biology* 37 (3): 515–555.

Jones, C. D., J. R. Troy, and L. Y. Pomara. 2007. "Similarities between *Campephilus* Woodpecker Double Raps and Mechanical Sounds Produced by Duck Flocks." *Wilson Journal of Ornithology* 119 (2): 259–262.

Jones, L. 1905. "Some April and May Work Suggested." *Wilson Bulletin* 17:22.

Joos, M. 1948. *Acoustic Phonetics*. Baltimore: Linguistic Society of America.

Judd, S. 1899. "Graphophone Demonstration of a Brown Thrasher's song."*Auk* 16 (1): 52–53.

Junker, T. 2003. "Ornithology and the Genesis of the Synthetic Theory of Evolution." *Avian Science* 3 (2/3): 65–73.

Kahn, D. 2002. "Concerning the Line." In *From Energy to Information: Representation in Science and Technology, Art, and Literature*, ed. B. Clarke and L. Dalrymple Henderson, 178–194. Stanford, CA: Stanford University Press.

Kaiser, D. 2005a. *Drawing Theories Apart: The Dispersion of Feynman Diagrams in Postwar Physics*. Chicago: University of Chicago Press.

Kaiser, D., ed. 2005b. *Pedagogy and the Practice of Science: Historical and Contemporary Perspectives*. Cambridge, MA: MIT Press.

Kaplan, J. 2013. "'Voices of the People': Linguistic research among Germany's prisoners of war during World War I." *Journal of the History of the Behavioral Sciences* 49 (3): 281–305.

Katz, M. 2004. *Capturing Sound: How Technology Has Changed Music*. Berkeley: University of California Press.

Keith, S. 1967. "Review of *Birds of North America: A Guide to Field Identification*." *Wilson Bulletin* 79 (2): 251–254.

Kellogg, E. W. 1955. "History of Sound Motion Pictures." *Journal of the Society for Motion Picture and Television Engineers* 64:291–304.

Kellogg, P. P. 1938. "A Study of Bird Sound Recording." Doctoral dissertation, Cornell University.

Kellogg, P. P. 1962a. "Bird-Sound Studies at Cornell." *Living Bird: Annual of the Cornell Laboratory of Ornithology* 1:37–48.

Kellogg, P. P. 1962b. "Vocalizations of the Black Rail (*Laterallus jamaicensis*) and the Yellow Rail (*Coturnicops noveboracensis*)." *Auk* 79 (4): 698–701.

Kellogg, P. P. 1962c. *A Field Guide to Western Bird Songs*. Boston: Houghton Mifflin. Three 33⅓ rpm discs.

Kellogg, P. P. 1965. *Bird Sounds of Marsh, Upland and Shore*. Washington, DC: National Geographic Society. Book and six 33⅓ rpm discs.

Kellogg, P. P. 1969. *Florida Bird Songs*. Ithaca, NY: Cornell University Laboratory of Ornithology. 33⅓ rpm disc.

Kellogg, P. P., and A. A. Allen. 1942. *American Bird Songs*. Vol. 1. Ithaca, NY: Comstock. Six 78 rpm discs.

Kellogg, P. P., and A. A. Allen. 1950. *Jungle Sounds*. Ithaca, NY: Comstock. 78 rpm disc.

Kellogg, P. P., and A. A. Allen. 1951. *American Bird Songs*. Vol. 2. Ithaca, NY: Comstock. Five 78 rpm discs.

Kellogg, P. P., and A. A. Allen. 1952. *Florida Bird Songs*. Ithaca, NY: Cornell University Press. 78 rpm disc.

Kellogg, P. P., and A. A. Allen. 1958. *An Evening in Sapsucker Woods*. Ithaca, NY: Cornell University Press. 33⅓ rpm disc.

Kellogg, P. P., and A. A. Allen. 1959. *A Field Guide to Bird Songs of Eastern and Central North America*. Boston: Houghton Mifflin. Two 33⅓ rpm discs.

Kellogg, P. P., and A. A. Allen. 1960. *Dawn in a Duckblind*. Boston: Houghton Mifflin. CH1089. 33⅓ rpm disc.

Kellogg, P. P., and J. H. Fassett. 1953. *Music and Bird Songs*. Ithaca, NY: Cornell University Laboratory of Ornithology. 33⅓ rpm disc.

Kellogg, P. P., and R. C. Stein. 1953. "Audio-Spectrographic Analysis of the Songs of the Alder Flycatcher." *Wilson Bulletin* 65 (2): 75–80.

Kevles, D. J. 1992. "Huxley and the Popularization of Science." In *Julian Huxley: Biologist and Statesman of Science*, ed. K. C. Waters and A. V. Helden, 238–251. Houston, TX: Rice University Press.

Kirby, D. 2011. *Lab Coats in Hollywood: Science, Scientists, and Cinema*. Cambridge, MA: MIT Press.

Kirkman, F. B. 1910. "The Meaning of Birds' Songs." *British Birds* 3 (4): 121.

Kivy, P. 1959. "Charles Darwin on Music." *Journal of the American Musicological Society* 12 (1): 42–48.

Klein, U. 2001. "Paper Tools in Experimental Cultures." *Studies in History and Philosophy of Science* 32 (2): 265–302.

Knorr Cetina, K. 1992. "The Couch, the Cathedral, and the Laboratory: On the Relationship between Experiment and Laboratory in Science." In *Science as Practice and Culture*, ed. A. Pickering, 113–138. Chicago: University of Chicago Press.

Knorr Cetina, K. 1999. *Epistemic Cultures: How the Sciences Make Knowledge*. Cambridge, MA: Harvard University Press.

Koch, L. 1955. *Memoirs of a Birdman*. Boston: Charles T. Branford.

Koenig, W., H. K. Dunn, and L. Y. Lacy. 1946. "The Sound Spectrograph." *Journal of the Acoustical Society of America* 15 (1): 19–49.

Kohler, R. E. 1994. *Lords of the Fly: Drosophila Genetics and the Experimental Life.* Chicago: University of Chicago Press.

Kohler, R. E. 2002a. *Landscapes and Labscapes: Exploring the Lab-Field Border in Biology.* Chicago: University of Chicago Press.

Kohler, R. E. 2002b. "Place and Practice in Field Biology." *History of Science* 40 (2): 189–210.

Kohler, R. E. 2006. *All Creatures: Naturalists, Collectors, and Biodiversity, 1850–1950.* Princeton, NJ: Princeton University Press.

Kohlstedt, S. G. 2010. *Teaching Children Science: Hands-On Nature Study in North America, 1890–1930.* Chicago: University of Chicago Press.

Konishi, M. 1964. "Song Variation in a Population of Oregon Juncos." *Condor* 66 (5): 423–436.

Krebs, S. 2012. "'Sobbing, Whining, Rumbling': Listening to Automobiles as Social Practice." In *The Oxford Handbook of Sound Studies*, ed. T. Pinch and K. Bijsterveld, 79–101. Oxford: Oxford University Press.

Kroodsma, D. E. 2007. *The Singing Life of Birds: The Art and Science of Listening to Birdsong.* New York: Houghton Mifflin.

Kroodsma, D. E., G. F. Budney, R. W. Grotke, J. Vielliard, R. Ranft, and O. D. Veprintseva. 1996. "Natural Sound Archives." In *Ecology and Evolution of Acoustic Communication among Birds*, ed. D. E. Kroodsma and E. H. Miller, 474–486. New York: Cornell University Press.

Kroodsma, D. E., and E. H. Miller. 1980. "Winter Wren Singing Behavior: A Pinnacle of Song Complexity." *Condor* 82 (4): 357–365.

Kuklick, H., and R. E. Kohler, eds. 1996. "Science in the Field." *Osiris* 11:1–14.

Kursell, J. 2003. *Schallkunst: Eine Literaturgeschichte der Musik in der Frühen Russischen Avantgarde.* Munich: Wiener Slawistischer Almanach.

Kursell, J. 2012. "A Gray Box: The Phonograph in Laboratory Experiments and Fieldwork, 1900–1920." In *The Oxford Handbook of Sound Studies*, ed. T. Pinch and K. Bijsterveld, 176–200. Oxford: Oxford University Press.

Kwa, C. 2008. "Painting and Photographing Landscapes: Pictorial Conventions and Gestalts.' *Configurations* 16 (1): 57–75.

Lachmund, J. 1999. "Making Sense of Sound: Auscultation and Lung Sound Codification in Nineteenth-Century French and German Medicine." *Science, Technology & Human Values* 24 (4): 419–450.

Lachmund, J. 2013. *Greening Berlin: The Co-production of Science, Politics, and Urban Nature.* Cambridge, MA: MIT Press.

Lachmund, J. 2015. "Strange birds: Ornithology and the Advent of the Collared Dove." *Science in Context* 28 (2): 259–284.

LaFollette, M. C. 2008. *Science on the Air: Popularizers and Personalities on Radio and Early Television.* Chicago: University of Chicago Press.

LaFollette, M. C. 2013. *Science on American Television: A History.* Chicago: University of Chicago Press.

Lancaster, D. A., and J. R. Johnson, eds. 1976. *Newsletter to Members 80.* Ithaca, NY: Cornell University Laboratory of Ornithology.

Landois, H. 1847. *Thierstimmen.* Freiburg in Bresgau: Herdersche Verlagsbuchhandlung.

Lange, B. 2015. "*Poste Restante*, and Messages in Bottles: Sound Recordings of Indian Prisoners in World War I." *Social Dynamics* 41 (1): 84–100.

Lanyon, W., and W. Fish. 1958. "Geographical Variation in the Vocalization of the Western Meadowlark." *Condor* 60 (5): 339–341.

Lanyon, W., and W. N. Tavolga, eds. 1960. *Animal Sounds and Communication.* Washington, DC: American Institute of Biological Sciences.

Latour, B. 1986. "Visualization and Cognition: Thinking with Eyes and Hands." *Knowledge and Society: Studies in the Sociology of Culture Past and Present* 6 (1): 1–40.

Latour, B. 1995. "The 'Pedofil' of Boa Vista: A Photo-Philosophical Montage." *Common Knowledge* 4 (1): 144–187.

Latour, B. 2000. "The Berlin Key or How to Do Things with Words." Trans. L. Davis. In *Matter, Materiality and Modern Culture*, ed. P. M. Graves-Brown, 10–21. London: Routledge.

Latour, B., and S. Woolgar. 1986. *Laboratory Life: The Construction of Scientific Facts.* Princeton, NJ: Princeton University Press.

Law, J., and M. Lynch. 1988. "Lists, Field Guides and the Descriptive Organization of Seeing: Birdwatching as an Exemplary Observational Activity." *Human Studies* 11 (2): 271–303.

Lawrence, A. 2006. "'No Personal Motive?' Volunteers, Biodiversity, and the False Dichotomies of Participation." *Ethics, Place and Environment: A Journal of Philosophy and Geography* 9 (3): 279–298.

Lawrence, A., and E. Turnhout. 2010. "Personal Meaning in the Public Sphere: The Standardisation and Rationalisation of Biodiversity Data in the UK and the Netherlands." *Journal of Rural Studies* 26 (4): 353–360.

Lawrence, C., and S. Shapin, eds. 1998. *Science Incarnate: Historical Embodiments of Natural Knowledge*. Chicago: University of Chicago Press.

Lemon, R. E. 1968. "The Relation between Organization and Function of Song in Cardinals." *Behaviour* 32 (1/3): 158–178.

Lemon, R. E., and C. Chatfield. 1971. "Organization of Song in Cardinals." *Animal Behaviour* 19 (1): 1–17.

Lemon, R. E., and A. Herzog. 1969. "The Vocal Behavior of Cardinals and Pyrrhuloxias in Texas." *Condor* 71 (1): 1–15.

Leonelli, S. 2011. "Packaging Small Facts for Re-use: Databases in Model Organism Biology." In *How Well Do Facts Travel? The Dissemination of Reliable Knowledge*, ed. P. Howlett and M. S. Morgan, 325–348. Cambridge: Cambridge University Press.

Leopold, A. 1936. "Threatened Species." In *The River of the Mother of God and Other Essays by Aldo Leopold*, ed. S. L. Flader and J. Baird Callicott, 230–234. Madison: University of Wisconsin Press.

Lesser, E. L. 2000. *Knowledge and Social Capital: Foundations and Applications*. Boston: Butterworth Heinemann.

Levin, D. M. 1993. *Modernity and the Hegemony of Vision*. Berkeley: University of California Press.

Lewis, D. 2012. *The Feathery Tribe: Robert Ridgway and the Modern Study of Birds*. New Haven, CT: Yale University Press.

Liberman, A. M., and F. S. Cooper. 1972. "In Search of the Acoustic Cues." *In Papers on Linguistics and Phonetics in Memory of Pierre Delattre*, ed. A. Valdman, 329–338. The Hague: Mouton.

Little, R. S. 2003. *For the Birds: The Laboratory of Ornithology and Sapsucker Woods at Cornell University*. Basking Ridge, NJ: Randolph S. Little.

Lutts, R. H. 1990. *The Nature Fakers: Wildlife, Science, and Sentiment*. Charlottesville: University Press of Virginia.

Lynch, M. 1985. "Discipline and the Material Form of Images: An Analysis of Scientific Visibility." *Social Studies of Science* 15 (1): 37–66.

Lynch, M. 1988. "The Externalized Retina: Selection and Mathematization in the Visual Documentation of Objects in the Life Sciences." *Human Studies* 11 (2/3): 201–234.

Lynch, M. 2011. "Credibility, Evidence, and Discovery: The Case of the Ivory-Billed Woodpecker." *Ethnographic Studies* 12 (1): 78–105.

Lynch, M., and S. Woolgar, eds. 1988. *Representations in Scientific Practice*. Cambridge, MA: MIT Press. First appeared as a special issue of *Human Studies* 11 (2/3).

MacDonald, H. 2002. "What Makes You a Scientist Is the Way You Look at Things: Ornithology and the Observer." *Studies in History and Philosophy of Biological and Biomedical Sciences* 33 (1): 53–77.

Mansell, J. 2017. *The Age of Noise in Britain: Hearing Modernity*. Urbana: University of Illinois Press.

Margoschis, R. 1977. *Recording Natural History Sounds*. Barnet, UK: Print and Press Services.

Marie, J. 2008. "For Science, Love and Money: The Social Worlds of Poultry and Rabbit Breeding in Britain, 1900–1940." *Social Studies of Science* 38 (6): 919–936.

Marler, P. 1952. "Variation in the Song of the Chaffinch 'Fringilla Coelebs.'" *Ibis* 94 (3): 458–472.

Marler, P. 1956. "The Voice of the Chaffinch and Its Function as a Language." *Ibis* 98 (2): 231–261.

Marler, P. 1959. "Developments in the Study of Animal Communication." In *Darwin's Biological Work*, ed. P. R. Bell, 150–206. Cambridge: Cambridge University Press.

Marler, P. 1960. "Bird Songs and Mate Selection." In *Animal Sounds and Communication*, ed. W. Lanyon and W. N. Tavolga, 349–367. Washington, DC: American Institute of Biological Sciences.

Marler, P. 1985. "Hark Ye to the Birds: Autobiographical Marginalia." In *Studying Animal Behavior: Autobiographies of the Founders*, ed. D. A. Dewsbury, 315–345. Chicago: University of Chicago Press.

Marler, P. 2004. "Science and Birdsong: The Good Old Days." In *Nature's Music: The Science of Birdsong*, ed. P. Marler and H. Slabbekoorn, 1–38. Amsterdam: Elsevier Academic Press.

Marler, P., and D. Isaac. 1960a. "Physical Analysis of a Simple Bird Song as Exemplified by the Chipping Sparrow." *Condor* 62 (2): 124–135.

Marler, P., and D. Isaac. 1960b. "Song Variation in a Population of Brown Towhees." *Condor* 62 (4): 272–283.

Marler, P., and H. Slabbekoorn, eds. 2004. *Nature's Music: The Science of Birdsong*. Amsterdam: Elsevier Academic Press.

Marler, P., M. Kreit, and M. Tamura. 1962. "Song Development in Hand-Raised Oregon Juncos." *Auk* 79 (1): 12–30.

Marler, P., and M. Tamura. 1962. "Song 'Dialects' in Three Populations of White-Crowned Sparrows." *Condor* 64 (5): 368–377.

Marshall, J. T., Jr. 1964. "Voice in Communication and Relationships among Brown Wowhees." *Condor* 66 (5): 345–356.

Marshall, J. T., Jr. 1977. "Audiospectrograms with Pitch Scale: A Universal "Language" for Representing Bird Songs Graphically." *Auk* 94 (1): 150–152.

Marten, K., and P. Marler. 1977. "Sound Transmission and Its Significance for Animal Temperate Habitats." *Behavioral Ecology and Sociobiology* 2 (3): 271–290.

Mason, C. R., P. P. Kellogg, and A. A. Allen. 1954. *The Mockingbird Sings.* Ithaca, NY: Cornell University Laboratory of Ornithology. 78 rpm disc.

Massinon, P. 2016. "Active Listening: The Cultural Politics of Magnetic Recording Technologies in North America, 1945–1993." Doctoral dissertation, University of Michigan.

Mathews, F. S. 1904. *Field Book of Wild Birds and Their Music: A Description of the Character and Music of Birds, Intended to Assist in the Identification of Species Common in the United States East of the Rocky Mountains.* New York: Putnam.

Matless, D. 2005. "Sonic Geography in a Nature Region." *Social & Cultural Geography* 6 (5): 745–766.

Mauss, M. 2006. *The Gift: The Form and Reason for Exchange in Archaic Societies.* London: Routledge.

Mauz, I., and C. Granjou. 2013. "A New Border Zone in Science: Collaboration and Tensions between Modeling Ecologists and Field Naturalists." *Science as Culture* 22 (3): 314–343.

Mayfield, H. 1957. "Proceedings of the Seventy-Fourth Stated Meeting of the American Ornithologists' Union." *Auk* 74 (1): 79–89.

Mayfield, H. 1979. Commentary: The Amateur in Ornithology. *Auk* 96 (1): 168–171.

McCain, K. W. 1991. "Communication, Competition, and Secrecy: The Production and Dissemination of Research-Related Information in Genetics." *Science, Technology & Human Values* 16 (4): 491–516.

McCray, W. P. 2000. "Large Telescopes and the Moral Economy of Recent Astronomy." *Social Studies of Science* 30 (5): 685–711.

Mearns, B., and R. Mearns. 1998. *The Bird Collectors.* London: Academic Press.

Merton, R. 1971. "The Normative Structure of Science." In *The Sociology of Science,* ed. N. W. Storer, 267–278. Chicago: University of Chicago Press.

Miller, E. H. 1980. "Commentary: Sound Spectrographic Analysis." *Condor* 82 (2): 234.

Mills, M. 2010. "Deaf jam: From Inscription to Reproduction to Information." *Social Text* 28 (1): 35–58.

Mills, M. 2018. *On the Phone: Deafness and Communication Engineering*. Durham, NC: Duke University Press.

Mirowski, P., and E.-M. Sent, eds. 2002. *Science Bought and Sold: Essays in the Economics of Science*. Chicago: University of Chicago Press.

Mitman, G. 1996. "'When Nature Is the Zoo': Vision and Power in the Art and Science of Natural History." *Osiris* 11:117–143.

Mitman, G. 1999. *Reel Nature: America's Romance with Wildlife on Film*. Cambridge, MA: Harvard University Press.

Mody, C. C. M. 2005. "The Sounds of Science: Listening to Laboratory Practice." *Science, Technology & Human Values* 30 (2): 175–198.

Mody, C. C. M. 2011. *Instrumental Community: Probe Microscopy and the Path to Nanotechnology*. Cambridge, MA: MIT Press.

Moore, R. T. 1913. "The Fox Sparrow as a Songster." *Auk* 30 (2): 177–187.

Moore, R. T. 1915. "Methods of Recording Bird Songs." *Auk* 32 (4): 535–538.

Moore, R. T. 1916. "Graphic Representation of Bird Song." *Auk* 33 (2): 228–230.

Moreau, R. E. 1959. "The Centenarian 'Ibis.'" *Ibis* 101 (1): 19–38.

Morton, D. 2000. *Off the Record: The Technology and Culture of Sound Recording in America*. New Brunswick, NJ: Rutgers University Press.

Morton, D. 2004. *Sound Recording: The Life Story of a Technology*. Westport, CT: Greenwood Press.

Morton, E. S. 1975. "Ecological Sources of Selection on Avian Sounds." *American Naturalist* 109 (965): 17–34.

Moss, S. 2004. *A Bird in the Bush: A Social History of Birdwatching*. London: Aurum.

Mundy, R. 2009. "Birdsong and the Image of Evolution." *Society & Animals* 17 (3): 206–223.

Mundy, R. 2010. "Nature's Music: Birds, Beasts, and Evolutionary Listening in the Twentieth Century." Doctoral dissertation, New York University.

Mundy, R. Forthcoming. *Animal Musicalities: Birds, Beasts, and Evolution in the Twentieth Century*. Middletown, CT: Wesleyan University Press.

Murie, O. J. 1962. "Why Do Birds Sing?" *Wilson Bulletin* 74 (2): 177–182.

Murphy, M. 2006. *Sick Building Syndrome and the Problem of Uncertainty: Environmental Politics, Technoscience, and Women Workers*. Durham, NC: Duke University Press.

Myers, N. 2008. "Molecular Embodiments and the Body-Work of Modeling in Protein Crystallography." *Social Studies of Science* 38 (2): 163–199.

Nelkin, D. 1984. *Science as Intellectual Property.* Washington, DC: American Association for the Advancement of Science.

Nelson, D. A., and P. Marler. 1990. "The Perception of Birdsong and an Ecological Concept of Signal Space." In *Comparative Perception, Vol. 2: Complex Signals,* ed. M. A. Berkley and W. C. Stebbins, 443–478. New York: Wiley.

Nemeth, E., and H. Brumm. 2010. "Birds and Anthropogenic Noise: Are Urban Songs Adaptive?" *American Naturalist* 176 (4): 465–475.

Nettl, B. 2015. *The Study of Ethnomusicology: Thirty-Three Discussions.* 3rd ed. Urbana: University of Illinois Press.

Nettl, B., and P. V. Bohlman, eds. 1991. *Comparative Musicology and Anthropology of Music.* Chicago: University of Chicago Press.

Neumann, J. 2005. "Ein Gewisser Cornel Schmitt." *Studienarchiv Umweltgeschichte* 10:15–18.

Nicholson, E. M. 1974. "British Library of Wildlife Sounds, Fifth Anniversary Address." *Recorded Sound* 55/56 (July–October): 330–333.

Nicholson, E. M., and L. Koch. 1936. *Songs of Wild Birds.* London: H. F. and G. Witherby.

Nicholson, E. M., and L. Koch. 1937. *More Songs of Wild Birds.* London: H. F. and G. Witherby.

Niethammer, G. 1955. "Jagd auf Vogelstimmen." *Journal für Ornithologie* 96 (1): 115–158.

Nikolsky, I. D. 1973. *Bioacoustics in the Service of Progress* [in Russian]. Moscow: Znaniye Publishing House.

Nikolsky, I. D. 1975. *Catalog of Recordings in the Library of Animal Voice, Department of Vertebrate Zoology, Moscow State University* [in Russian]. Moscow: VINITI.

North, M. E. W. 1950. "Transcribing Bird-Song." *Ibis* 92 (1): 99–114.

North, M. E. W., and D. McChesney. 1964. *More Voices of African Birds.* Ithaca, NY: Cornell Laboratory of Ornithology and Houghton Mifflin. CH1101. 33⅓ rpm disc.

Nyhart, L. K. 1991. "Writing Zoologically: The *Zeitschrift für wissenschaftliche Zoologie* and the Zoological Community in Late Nineteenth-Century Germany." In *The Literary Structure of Scientific Argument: Historical Studies,* ed. P. Dear, 43–71. Philadelphia: University of Pennsylvania Press.

Nyhart, L. K. 1998. "Civic and Economic Zoology in Nineteenth-Century Germany: The 'Living Communities' of Karl Mobius." *Isis* 89 (4): 605–630.

Nyhart, L. K. 2009. *Modern Nature: The Rise of the Biological Perspective in Germany.* Chicago: University of Chicago Press.

Ohala, J. J. 2014. "Speech Technology: Historical Antecedents." In *Concise History of the Language Sciences: From the Sumerians to the Cognitivists,* ed. E. F. K. Koerner and R. E. Asher, 416–419. Oxford: Pergamon.

Olds, W. B. 1914. *Twenty-Five Bird Songs for Children.* London: G. Schirmer.

Olds, W. B. 1916. *Bird Songs, Flower Songs Gamble Hinged Music.* Chicago, IL: Gamble Hinged Music.

Oldys, H. 1904a. Lectures on birds by Henry Oldys. n.p.

Oldys, H. 1904b. "The Rhythmical Song of the Wood Pewee." *Auk* 212:270–274.

Oldys, H. 1913. "A Remarkable Hermit Thrush Song." *Auk* 30 (4): 538–541.

Oldys, H. 1916. "Rhythmical Singing of Veeries." *Auk* 33 (1): 17–21.

Orr, O. H., Jr. 1992. *Saving American Birds: T. Gilbert Pearson and the Founding of the Audubon Movement.* Gainesville: University Press of Florida.

Pantalony, D. 2009. *Altered Sensations: Rudolph Koenig's Acoustical Workshop in Nineteenth-Century Paris.* Dordrecht: Springer.

Parker, J. N., and E. J. Hackett. 2014. "The Sociology of Science and Emotions." In *Handbook of the Sociology of Emotions,* ed. J. E. Stets and J. H. Turner, 549–572. Dordrecht: Springer.

Parr, J. 2010. *Sensing Changes: Technologies, Environments, and the Everyday, 1953–2003.* Vancouver: UBC Press.

Perlman, M. 2004. "Golden Ears and Meter Readers." *Social Studies of Science* 34 (5): 783–807.

Pesic, Peter. 2014. *Music and the Making of Modern Science.* Cambridge, MA: MIT Press.

Peterson, R. T. [1934] 1965. *A Field Guide to the Birds.* Boston: Houghton Mifflin.

Pinch, T., and K. Bijsterveld. 2012. "New Keys to the World of Sound." In *The Oxford Handbook of Sound Studies,* ed. T. Pinch and K. Bijsterveld, 3–35. Oxford: Oxford University Press.

Pinch, T., and F. Trocco. 2002. *Analog Days: The Invention and Impact of the Moog Synthesizer.* Cambridge, MA: Harvard University Press.

Porcello, T. 2004. "Speaking of Sound: Language and the Professionalization of Sound-Recording Engineers." *Social Studies of Science* 34 (5): 733–758.

Potter, R. K. 1945. "Visible Patterns of Sound." *Science* 102 (2654): 463–470.

Potter, R. K., G. A. Kopp, and H. C. Green. 1947. *Visible Speech*. London: Macmillan.

Poulsen, H. 1951. "Inheritance and Learning in the Song of the Chaffinch." *Behaviour* 3 (3): 216–228.

Pryer, A. 2002. "Graphic Notation." In *The Oxford Companion to Music*, ed. A. Latham, 537. Oxford: Oxford University Press.

Pulling, M. J. L. 1949. "The Development of Mobile Recording technique." *BBC Quarterly* 4 (3): 179–192.

Purves, F. 1962. *Bird Song Recording*. London: Focal Press.

Radick, G. 2007. *The Simian Tongue: The Long Debate about Animal Language*. Chicago: University of Chicago Press.

Ramseyer, U. J. 1908. *Unsere Singvögel: Ihr Gesang, Leben und Lieben*. Narau: Emil Wirz vorm. J. J. Chriften.

Randel, D. M. 2003. *The Harvard Dictionary of Music*. Cambridge, MA: Harvard University Press.

Ranft, R. 2004. "Natural Sound Archives: Past, Present and Future." *Annals of the Brazilian Academy of Sciences* 76 (4): 455–465.

Rasmussen, N. 2004. "The Moral Economy of the Drug Company–Medical Scientist Collaboration in Interwar America." *Social Studies of Science* 34 (2): 161–185.

Read, O., and W. Welch. 1976. *From Tin Foil to Stereo: Evolution of the Phonograph*. Indianapolis, IN: Bobbs-Merrill.

Rees, A. 2005. "A Place That Answers Questions: Primatological Field Sites and the Making of Authentic Observations." *Studies in History and Philosophy of Biological and Biomedical Sciences* 37 (2): 311–333.

"Reflectors like Airplane Detectors Catch Bird Songs." 1934. *Science News Letter*, November 10, 299–300.

Rehding, A. 2000. "The Quest for the Origins of Music in Germany circa 1900." *Journal of the American Musicological Society* 53 (2): 345–385.

Reynard, G. B., and P. P. Kellogg. 1969. *Caribbean Bird Songs: 54 Species in Puerto Rico and the Virgin Islands*. Ithaca, NY: Cornell Laboratory of Ornithology. 33⅓ rpm disc.

Rice, T. 2008. "'Beautiful murmurs': Stethoscopic Listening and Acoustic Objectification." *Senses and Society* 3 (3): 293–306.

Ridgway, R. 1901. "The Birds of North and Middle America: A Descriptive Catalogue, Pt. I." *Bulletin of the United States National Museum* 50.

Ritvo, H. 1989. *The Animal Estate: The English and Other Creatures in the Victorian Age.* Cambridge, MA: Harvard University Press.

Robbins, C. S., B. Bruun, and H. S. Zim. 1966. *Birds of North America: A Guide to Field Identification.* New York: Golden Press.

Roberts, L. 1995. "The Death of the Sensuous Chemist: The "New" Chemistry and the Transformation of Sensuous Technology." *Studies in History and Philosophy of Science* 26 (4): 503–529.

Roth, W.-M. 2003. *Toward an Anthropology of Graphing: Semiotic and Activity-Theoretic Perspectives.* Dordrecht: Kluwer Academic Publishers.

Roth, W.-M., and G. M. Bowen. 1999. "Digitizing Lizards: The Topology of "Vision" in Ecological Fieldwork." *Social Studies of Science* 29 (5): 719–764.

Rothenberg, D. 2005. *Why Birds Sing: A Journey into the Mystery of Birdsong.* New York: Basic Books.

Rowan, W. M. 1925. "A Practical Method of Recording Bird-Calls." *British Birds* 18 (1): 14–18.

Rudwick, M. 1976. "The Emergence of a Visual Language for Geological Science, 1760–1840." *History of Science* 14:149–195.

Sage, J. H. 1899. "Sixteenth Congress of the A. O. U." *Auk* 16 (1): 51–55.

Sauer, T. 2009. *Notations 21.* New York: Mark Batty Publisher.

Saunders, A. A. 1915. "Some Suggestions for Better Methods of Recording and Studying Bird Songs." *Auk* 32 (2): 173–183.

Saunders, A. A. 1916a. Correspondence "Graphic Representation of Bird Song." *Auk* 33 (2): 228–230.

Saunders, A. A. 1916b. "Methods of Recording Bird Song." *Auk* 33 (1): 103–107.

Saunders, A. A. 1922. "The Song of the Field Sparrow." *Auk* 39 (3): 386–399.

Saunders, A. A. 1924. "Recognizing Individual Birds by Song." *Auk* 41 (2): 242–259.

Saunders, A. A. 1961. "The Songs and Calls of the Wood Thrush." *Auk* 78 (4): 595–606.

Saunders, W. E. 1934. "Losing the Bird Songs." *Auk* 51 (4): 503–556.

Schafer, R. M. 1977. *The Soundscape: Our Sonic Environment and the Tuning of the World.* Rochester, VT: Destiny Books.

Schaffer, S. 1988. "Astronomers Mark Time: Discipline and the Personal Equation." *Science in Context* 2 (1): 115–145.

Schaffner, S. 2011. *Binocular Vision: The Politics of Representation in Birdwatching Field Guides*. Amherst and Boston: University of Massachusetts Press.

Schalow, H. 1911. *Verhandlungen des V. Internationalen Ornithologen-Kongresses in Berlin*. Berlin: Deutsche Ornithologische Gesellschaft.

Schmidgen, H. 2003. "Time and Noise: The Stable Surroundings of Reaction Experiments, 1860–1890." *Studies in History and Philosophy of Biological and Biomedical Sciences* 34 (2): 237–275.

Schmidgen, H. 2008. "Silence in the Laboratory: The History of Soundproof Rooms." In *Sounds of Science—Schall im Labor*, ed. J. Kursell, 47–62. Berlin: Max Planck Institute for the History of Science.

Schmidt Horning, S. 2013. *Chasing Sound: Technology, Culture, and the Art of Studio Recording from Edison to the LP*. Baltimore: Johns Hopkins University Press.

Schmitt, C. 1923. *Zwiesprache mit der Natur*. Freising: Verlag Dr. F. P. Datterer and Cie.

Schmitt, C. 1932. *Die Stimme der Natur, Wege zur Naturliebe*. Vol. 10. Freising: Verlag Dr. F. P. Datterer and Cie.

Schmitt, C., and H. Stadler. 1913. "Studien über Vogelstimmen." *Journal für Ornithologie* 61 (2): 383–394.

Schmitt, C., and H. Stadler. 1919. *Die Vogelsprache: Eine Anleitung zu ihrer Erkennung und Erforschung*. Stuttgart: Franckh.

Schmitt, P. J. [1969] 1990. *Back to Nature: The Arcadian Myth in Urban America*. Baltimore: Johns Hopkins University Press.

Schwartz, P. 1963. *Bird Songs from the Tropics*. Caracas: Instituto Neotropical. 33⅓ rpm disc.

Secord, A. 2002. "Botany on a Plate: Pleasure and the Power of Pictures in Promoting Early Nineteenth-Century Scientific Knowledge." *Isis* 93 (1): 28–57.

Secord, J. A. 2004. "Knowledge in Transit." *Isis* 95 (4): 654–672.

Sellar, P. J. 1976. "Sound Recording and the Birdwatcher." *British Birds* 69 (6): 202–215.

Shapin, S. 1994. *A Social History of Truth: Civility and Science in Seventeenth-Century England*. Chicago: University of Chicago Press.

Shapin, S. 1998. "Placing the View from Nowhere: Historical and Sociological Problems in the Location of Science." *Transactions of the Institute of British Geographers* 23 (1): 5–12.

Shelemay, K. K. 1991. "Recording Technology, the Record Industry, and Ethnomusicological Scholarship." In *Comparative Musicology and Anthropology of Music: Essays on the History of Ethnomusicology*, ed. B. Nettl, 277–292. Chicago: University of Chicago Press.

Shiovitz, K. A. 1975. "The Process of Species-Specific Song Recognition by the Indigo Bunting, *Passerina cyanea*, and Its Relationship to the Organization of Avian Acoustical Behavior." *Behaviour* 55 (1/2): 128–179.

Simms, E. 1976. *Birds of the Air: The Autobiography of a Naturalist and Broadcaster*. London: Hutchinson.

Simms, E., and G. F. Wade. 1953. "Recent Advances in the Recording of Bird-Songs." *British Birds* 46 (6): 200–209.

Slabbekoorn, H. 2009. "Geschiedenis van de Bioakoestiek in technologisch perspectief." *Tijdschrift voor Mediageschiedenis* 12 (2): 205–232.

Slabbekoorn, H., and M. Peet. 2003. "Birds Sing at a Higher Pitch in Urban Noise." *Nature* 424 (6946): 267.

Slabbekoorn, H., Y. Xiao-Jing, and W. Halfwerk. 2012. "Birds and Anthropogenic Noise: Singing Higher May Matter. (A Comment on Nemeth and Brumm, 'Birds and Anthropogenic Noise: Are Urban Songs Adaptive?')." *American Naturalist* 180 (1): 142–145.

Slater, P. J. B., and S. A. Ince. 1979. "Cultural Evolution in Chaffinch Song." *Behaviour* 71 (1/2): 146–166.

Smith, J. 2015. *Eco-Sonic Media*. Oakland: University of California Press.

Smith, P. H. 2004. *The Body of the Artisan: Art and Experience in the Scientific Revolution*. Chicago: University of Chicago Press.

Stadler, H., and C. Schmitt. 1914. "The Study of Bird-Notes." *British Birds* 8 (1): 2–8.

Stangl, B. 2000. *Ethnologie im Ohr: Die Wirkungsgeschichte des Phonographen*. Vienna: WUV Universitätsverlag.

Star, S. L., and J. R. Griesemer. 1989. "Institutional Ecology, "Translations" and Boundary Objects: Amateurs and Professionals in Berkeley's Museum of Vertebrate Zoology, 1907–39." *Social Studies of Science* 19 (3): 387–420.

Stein, R. C. 1956. "A Comparative Study of 'Advertising Song' in the *Hylocichla* Thrushes." *Auk* 73 (4): 503–512.

Steinberg, M. K. 2008. *Stalking the Ghost Bird: The Elusive Ivory-Billed Woodpecker in Louisiana*. Baton Rouge: Louisiana State University Press.

Sterne, J. 2003. *The Audible Past: Cultural Origins of Sound Reproduction.* Durham, NC: Duke University Press.

Sterne, J. 2012. *MP3: The Meaning of a Format.* Durham, NC: Duke University Press.

Stillwell, J., and N. Stillwell. 1961. *Bird Songs of Dooryard, Field and Forest.* Vol. 1. Old Greenwich, CT: Ficker Recording Service. C101:33. 33⅓ rpm disc.

Stillwell, N. 1964. *Bird Song: Adventures and Techniques in Recording the Songs of American Birds.* New York: Doubleday.

Stone, W. 1913. "Notes and News." *Auk* 30 (3): 472–474.

Stonehouse, B. 1963. "Review of *Bird Song: The Biology of Vocal Communication and Expression in Birds.*" *Science Progress* 51 (201): 155–156.

Strasser, B. B. J. 2011. "The Experimenter's Museum: GenBank, Natural History, and the Moral Economies of Biomedicine." *Isis* 102 (1): 60–96.

Stratton-Porter, G. 1903. *The Song of the Cardinal: A Love Story.* Indianapolis, IN: Bobbs-Merrill.

Stratton-Porter, G. 1910. *Music of the Wild: With Reproductions of the Performers, Their Instruments and Festival Halls.* New York: Eaton and Mains.

Strong, R. M. 1918. "The Description of the Voice of Birds." *Auk* 35 (2): 133–135.

Stubbs, F. J. 1910. "The Meaning of Birds' Songs." *British Birds* 3 (5): 155–157.

Stumpf, C. 1911. *Die Anfänge der Musik.* Leipzig: Johann Ambrosius Barth.

Summers, E. 1916. "Notation of Bird Songs and Notes." *Auk* 33 (1): 78–80.

Supper, A. 2012. "Lobbying for the Ear: The Sonification of Scientific Data and the Struggles for Its Academic Legitimacy." Doctoral dissertation, University of Maastricht.

Supper, A., and K. Bijsterveld. 2015. "Sounds Convincing: Modes of Listening and Sonic Skills in Knowledge Making." *Interdisciplinary Science Reviews* 40 (2): 124–144.

Szöke, P. 1987. *The Unknown Music of Birds.* Budapest: Hungaroton MHV. 33⅓ rpm disc.

Tapia, J. E. 1997. *Circuit Chautauqua: From Rural Education to Popular Entertainment in Early Twentieth Century America.* Jefferson, NC: McFarland.

Taruskin, R. 2005. *The Oxford History of Western Music,* Vol. 5: *The Late Twentieth Century.* Oxford: Oxford University Press.

Taylor, T. 2014. *Strange Sounds: Music, Technology and Culture.* London: Routledge.

Thielcke, G. 1961. "Ergebnisse der Vogelstimmen-Analyse." *Journal für Ornithologie* 102 (3): 285–300.

Thomas, M. 2005. "Are Animals just Noisy Machines? Louis Boutan and the Co-invention of Animal and Child Psychology in the French Third Republic." *Journal of the History of Biology* 38 (3): 425–460.

Thompson, E. 1995. "Machines, Music, and the Quest for Fidelity: Marketing the Edison Phonograph in America, 1877–1925." *Musical Quarterly* 79 (1): 131–171.

Thompson, E. 1997. "Dead Rooms and Live Wires: Harvard, Hollywood, and the Deconstruction of Architectural Acoustics, 1900–1930." *Isis* 88 (4): 597–626.

Thompson, E. 2002. *The Soundscape of Modernity: Architectural Acoustics and the Culture of Listening in America 1900–1933*. Cambridge, MA: MIT Press.

Thompson, W. L. 1970. "Song Variation in a Population of Indigo Buntings." *Auk* 87 (1): 58–71.

Thompson, W. L. 1979. "Suggestions for preparing audiospectrograms for publication." *Condor* 81 (2): 220–221.

Thorpe, W. H. 1951. "The Learning Abilities of Birds." *Ibis* 93 (2): 252–296.

Thorpe, W. H. 1954. "The Process of Song-Learning in the Chaffinch as Studied by Means of the Sound Spectrograph." *Nature* 173:465–469.

Thorpe, W. H. 1958. "The Learning of Song Patterns by Birds, with Especial Reference to the Songs of the Chaffinch 'Fringilla Coelebs.'" *Ibis* 100 (4): 535–570.

Thorpe, W. H. 1961a. *Bird-Song: The Biology of Vocal Communication and Expression in Birds*. Cambridge: Cambridge University Press.

Thorpe, W. H. 1961b. "The Songs of Some Families of the Passeriformes. II. The Songs of the Buntings (Emberizidae)." *Ibis* 103a (2): 246–259.

Thorpe, W. H. 1966. "Ritualization in Ontogeny. II. Ritualization in the Individual Development of Bird Song." *Philosophical Transactions of the Royal Society of London B: Biological Sciences* 251 (772): 351–358.

Thorpe, W. H. 1968. "Obituary: Myles Edward Wentworth North, 1908–1967." *Ibis* 110 (3): 364–365.

Thorpe, W. H. 1979. *The Origins and Rise of Ethology*. London: Heinemann.

Thorpe, W. H., J. Hall-Craggs, B. Hooker, T. Hooker, and R. E. Hutchison. 1972. "Duetting and Antiphonal Song in Birds: Its Extent and Significance." *Behaviour* 18 (suppl.): 1–197.

Thorpe, W. H., and R. A. Hinde. 1956. "An Inexpensive Type of Sound-Proof Room Suitable for Zoological Research." *Experimental Biology* 33 (4): 750–755.

Thorpe, W. H., and B. I. Lade. 1961. "The Songs of Some Families of the Passeriformes. I. Introduction: The Analysis of Bird Songs and Their Expression in Graphic Notation." *Ibis* 103a (2): 231–245.

Thorpe, W. H., and M. E. North. 1965. "Origin and Significance of the Power of Vocal Imitation: With Special Reference to the Antiphonal Singing of Birds. *Nature* 208 (5007): 219–222.

Thorpe, W. H., and M. E. North. 1966. "Vocal Imitation in the Tropical Bou-Bou Shrike *Laniarius Aethopocus major* as a Means of Establishing and Maintaining Social Bonds." *Ibis* 108:432–435.

Tick, J., and P. Beaudoin, eds. 2008. *Music in the USA: A Documentary Companion.* Oxford: Oxford University Press.

Tillmann, H. G. 1995. "Early Modern Instrumental Phonetics." In *Concise History of the Language Sciences: From the Sumerians to the Cognitivists*, ed. E. F. K. Koerner and R. E. Asher, 401–416. Oxford: Pergamon.

Tipp, C. 2011a. "Listening to the Nightingale Song." *Brio: Journal of the IAML* 48 (2): 15–21.

Tipp, C. 2011b. "Margaret McKee: The California Mocking Bird." *Record* 39 (September): 392–396.

Tipp, C. 2011c. "An Overview of Early Commercial Wildlife Recordings at the British Library." *IASA Journal* 37 (July): 47–54.

Tkaczyk, V. 2014. Listening in circles: Spoken drama and the architects of sound, 1750–1830. *Annals of Science* 71 (3): 299–334.

Toogood, M. 2011. "Modern Observations: New Ornithology and the Science of Ourselves, 1920–1940." *Journal of Historical Geography* 37 (3): 348–357.

Torrey, B. 1885. *Birds in the Bush.* Boston: Houghton Mifflin.

Turchetti, S., N. Herran, and S. Boudia. 2012. "Introduction: Have We Ever Been 'Transnational'? Towards a History of Science across and beyond Borders." *British Journal for the History of Science* 45 (3): 319–336.

Van der Vleuten, E., and A. Kaijser, eds. 2006. *Networking Europe: Transnational Infrastructures and the Shaping of Europe, 1850–2000.* Sagamore Beach, MA: Science History Publications.

Veprintsev, B. N. 1980. "Wildlife Sound Recording in the Soviet Union." *Comparative Biochemistry and Physiology* 67 (A): 321–328.

Vetter, J. 2004. "Science along the Railroad: Expanding Field Work in the US Central West." *Annals of Science* 61 (2): 187–211.

Vetter, J. 2011a. "Introduction: Lay Observation in the History of Scientific Observation." *Science in Context* 24 (2): 127–141.

Vetter, J., ed. 2011b. *Knowing Global Environments: New Historical Perspectives on the Field Sciences.* New Brunswick, NJ: Rutgers University Press.

"Visible Bird Song." 1953. *Time Magazine*, July 20.

"Voices of Birds Record: Phonographic Reproductions Made at Stuttgart Zoo." 1933. *Herald Tribune*, September 24.

Voices of Southern Birds Recorded for Movie Films." 1935. *The Science Newsletter* 27 (736): 316.

Voigt, A. 1913. *Excursionsbuch zum Studium der Vogelstimmen: Praktische Anleitung zum Bestimmen der Vögel nach ihrem Gesange von Prof. Dr. A Voigt.* 6th ed. Leipzig: Quelle u. Meyer.

Volmar, A. 2015. *Klang-Experimente: Die Auditive Kultur der Naturwissenschaften 1761–1961.* Frankfurt am Main: Campus Verlag.

von Glahn, D. 2013. *Music and the Skillful Listener: American Women Compose the Natural World.* Bloomington: Indiana University Press.

Wachelder, J. 2009. "Vogelafbeeldingen en Vogels Kijken in de Negentiende en Twintigste Eeuw." *Tijdschrift voor Mediageschiedenis* 12 (2): 187–204.

Wallaschek, R. 1893. *Primitive Music.* New York: Longmans, Green.

Walters, J. R., and E. L. Crist. 2005. "Rediscovering the King of Woodpeckers: Exploring the Implications." *Avian Conservation & Ecology* 1 (1): 6–12.

Warwick, A. 2003. *Masters of Theory: Cambridge and the Rise of Mathematical Physics.* Chicago: University of Chicago Press.

Wheeler, W. C., and J. T. Nichols. 1924. "The Song of the Song Sparrow." *Auk* 41 (3): 444–451.

Williams, J. J. 1902. "A study of Bird Songs. Chapter I. Preparatory Study of Bird Songs." *Condor* 4 (1): 12–14.

Winn, W. 2009. "'Proof' in Pictures: Visual Evidence and Meaning Making in the Ivory-Billed Woodpecker Controversy." *Journal of Technical Writing and Communication* 39 (4): 351–379.

Witchell, C. A. 1896. *The Evolution of Bird-Song with Observations on the Influence of Heredity and Imitation.* London: Adam and Charles Black.

Wittje, R. 2016. *The Age of Electroacoustics: Transforming Science and Sound.* Cambridge, MA: MIT Press.

Woodward, A. 1923. *Whistling as an Art.* New York: Fischer.

Wouters, P., and P. Schröder, eds. 2003. *Promise and Practice in Data Sharing: The Public Domain of Digital Research Data*. Amsterdam: NIWI-KNAW.

Zimmerman, A. S. 2008. "New Knowledge from Old Data: The Role of Standards in the Sharing and Reuse of Ecological Data." *Science, Technology, & Human Values* 33 (5): 631–652.

Zon, B. 2007. *Representing Non-Western Music in Nineteenth-Century Britain*. Rochester, NY: University of Rochester Press.

Zuckerman, H. A. 1988. "Intellectual Property and Diverse Rights of Ownership in Science." *Science, Technology & Human Values* 13 (1/2): 7–16.

Index

Inside Technology

edited by Wiebe E. Bijker, W. Bernard Carlson, and Trevor Pinch

Joeri Bruyninckx, *Listening in the Field: Recording and the Science of Birdsong*

Edward Jones-Imhotep, *The Unreliable Nation: Hostile Nature and Technological Failure in the Cold War*

Jennifer L. Lieberman, *Power Lines: Electricity in American Life and Letters, 1882–1952*

Pablo J. Boczkowski and C. W. Anderson, editors, *Remaking the News: Essays on the Future of Journalism Scholarship in the Digital Age*

Benoît Godîn, *Models of Innovation: History of an Idea*

Brice Laurent, *Democratic Experiments: Problematizing Nanotechnology and Democracy in Europe and the United States*

Stephen Hilgartner, *Reordering Life: Knowledge and Control in the Genomics Revolution*

Cyrus C. M. Mody, *The Long Arm of Moore's Law: Microelectronics and American Science*

Harry Collins, Robert Evans, and Christopher Higgins, *Bad Call: Technology's Attack on Referees and Umpires and How to Fix It*

Tiago Saraiva, *Fascist Pigs: Technoscientific Organisms and the History of Fascism*

Teun Zuiderent-Jerak, *Situated Intervention: Sociological Experiments in Health Care*

Basile Zimmermann, *Technology and Cultural Difference: Electronic Music Devices, Social Networking Sites, and Computer Encodings in Contemporary China*

Andrew J. Nelson, *The Sound of Innovation: Stanford and the Computer Music Revolution*

Sonja D. Schmid, *Producing Power: The Pre-Chernobyl History of the Soviet Nuclear Industry*

Casey O'Donnell, *Developer's Dilemma: The Secret World of Videogame Creators*

Christina Dunbar-Hester, *Low Power to the People: Pirates, Protest, and Politics in FM Radio Activism*

Eden Medina, Ivan da Costa Marques, and Christina Holmes, editors, *Beyond Imported Magic: Essays on Science, Technology, and Society in Latin America*

Anique Hommels, Jessica Mesman, and Wiebe E. Bijker, editors, *Vulnerability in Technological Cultures: New Directions in Research and Governance*

Amit Prasad, *Imperial Technoscience: Transnational Histories of MRI in the United States, Britain, and India*

Charis Thompson, *Good Science: The Ethical Choreography of Stem Cell Research*

Tarleton Gillespie, Pablo J. Boczkowski, and Kirsten A. Foot, editors, *Media Technologies: Essays on Communication, Materiality, and Society*

Catelijne Coopmans, Janet Vertesi, Michael Lynch, and Steve Woolgar, editors, *Representation in Scientific Practice Revisited*

Rebecca Slayton, *Arguments That Count: Physics, Computing, and Missile Defense, 1949–2012*

Stathis Arapostathis and Graeme Gooday, *Patently Contestable: Electrical Technologies and Inventor Identities on Trial in Britain*

Jens Lachmund, *Greening Berlin: The Co-production of Science, Politics, and Urban Nature*

Chikako Takeshita, *The Global Biopolitics of the IUD: How Science Constructs Contraceptive Users and Women's Bodies*

Cyrus C. M. Mody, *Instrumental Community: Probe Microscopy and the Path to Nanotechnology*

Morana Alač, *Handling Digital Brains: A Laboratory Study of Multimodal Semiotic Interaction in the Age of Computers*

Gabrielle Hecht, editor, *Entangled Geographies: Empire and Technopolitics in the Global Cold War*

Michael E. Gorman, editor, *Trading Zones and Interactional Expertise: Creating New Kinds of Collaboration*

Matthias Gross, *Ignorance and Surprise: Science, Society, and Ecological Design*

Andrew Feenberg, *Between Reason and Experience: Essays in Technology and Modernity*

Wiebe E. Bijker, Roland Bal, and Ruud Hendricks, *The Paradox of Scientific Authority: The Role of Scientific Advice in Democracies*

Park Doing, *Velvet Revolution at the Synchrotron: Biology, Physics, and Change in Science*

Gabrielle Hecht, *The Radiance of France: Nuclear Power and National Identity after World War II*

Richard Rottenburg, *Far-Fetched Facts: A Parable of Development Aid*

Michel Callon, Pierre Lascoumes, and Yannick Barthe, *Acting in an Uncertain World: An Essay on Technical Democracy*

Ruth Oldenziel and Karin Zachmann, editors, *Cold War Kitchen: Americanization, Technology, and European Users*

Deborah G. Johnson and Jameson W. Wetmore, editors, *Technology and Society: Building Our Sociotechnical Future*

Trevor Pinch and Richard Swedberg, editors, *Living in a Material World: Economic Sociology Meets Science and Technology Studies*

Christopher R. Henke, *Cultivating Science, Harvesting Power: Science and Industrial Agriculture in California*

Helga Nowotny, *Insatiable Curiosity: Innovation in a Fragile Future*

Karin Bijsterveld, *Mechanical Sound: Technology, Culture, and Public Problems of Noise in the Twentieth Century*

Peter D. Norton, *Fighting Traffic: The Dawn of the Motor Age in the American City*

Joshua M. Greenberg, *From Betamax to Blockbuster: Video Stores and the Invention of Movies on Video*

Mikael Hård and Thomas J. Misa, editors, *Urban Machinery: Inside Modern European Cities*

Christine Hine, *Systematics as Cyberscience: Computers, Change, and Continuity in Science*

Wesley Shrum, Joel Genuth, and Ivan Chompalov, *Structures of Scientific Collaboration*

Shobita Parthasarathy, *Building Genetic Medicine: Breast Cancer, Technology, and the Comparative Politics of Health Care*

Kristen Haring, *Ham Radio's Technical Culture*

Atsushi Akera, *Calculating a Natural World: Scientists, Engineers and Computers during the Rise of US Cold War Research*

Donald MacKenzie, *An Engine, Not a Camera: How Financial Models Shape Markets*

Geoffrey C. Bowker, *Memory Practices in the Sciences*

Christophe Lécuyer, *Making Silicon Valley: Innovation and the Growth of High Tech, 1930–1970*

Anique Hommels, *Unbuilding Cities: Obduracy in Urban Sociotechnical Change*

David Kaiser, editor, *Pedagogy and the Practice of Science: Historical and Contemporary Perspectives*

Charis Thompson, *Making Parents: The Ontological Choreography of Reproductive Technology*

Pablo J. Boczkowski, *Digitizing the News: Innovation in Online Newspapers*

Dominique Vinck, editor, *Everyday Engineering: An Ethnography of Design and Innovation*

Nelly Oudshoorn and Trevor Pinch, editors, *How Users Matter: The Co-Construction of Users and Technology*

Peter Keating and Alberto Cambrosio, *Biomedical Platforms: Realigning the Normal and the Pathological in Late-Twentieth-Century Medicine*

Paul Rosen, *Framing Production: Technology, Culture, and Change in the British Bicycle Industry*

Maggie Mort, *Building the Trident Network: A Study of the Enrollment of People, Knowledge, and Machines*

Donald MacKenzie, *Mechanizing Proof: Computing, Risk, and Trust*

Geoffrey C. Bowker and Susan Leigh Star, *Sorting Things Out: Classification and Its Consequences*

Charles Bazerman, *The Languages of Edison's Light*

Janet Abbate, *Inventing the Internet*

Herbert Gottweis, *Governing Molecules: The Discursive Politics of Genetic Engineering in Europe and the United States*

Kathryn Henderson, *On Line and On Paper: Visual Representation, Visual Culture, and Computer Graphics in Design Engineering*

Susanne K. Schmidt and Raymund Werle, *Coordinating Technology: Studies in the International Standardization of Telecommunications*

Marc Berg, *Rationalizing Medical Work: Decision-Support Techniques and Medical Practices*

Eda Kranakis, *Constructing a Bridge: An Exploration of Engineering Culture, Design, and Research in Nineteenth-Century France and America*

Paul N. Edwards, *The Closed World: Computers and the Politics of Discourse in Cold War America*

Donald MacKenzie, *Knowing Machines: Essays on Technical Change*

Wiebe E. Bijker, *Of Bicycles, Bakelites, and Bulbs: Toward a Theory of Sociotechnical Change*

Louis L. Bucciarelli, *Designing Engineers*

Geoffrey C. Bowker, *Science on the Run: Information Management and Industrial Geophysics at Schlumberger, 1920–1940*

Wiebe E. Bijker and John Law, editors, *Shaping Technology / Building Society: Studies in Sociotechnical Change*

Stuart Blume, *Insight and Industry: On the Dynamics of Technological Change in Medicine*

Donald MacKenzie, *Inventing Accuracy: A Historical Sociology of Nuclear Missile Guidance*

Pamela E. Mack, *Viewing the Earth: The Social Construction of the Landsat Satellite System*

H. M. Collins, *Artificial Experts: Social Knowledge and Intelligent Machines*

http://mitpress.mit.edu/books/series/inside-technology

Printed in the United States
by Baker & Taylor Publisher Services